Green Energy and Technology

Climate change, environmental impact and the limited natural resources urge scientific research and novel technical solutions. The monograph series Green Energy and Technology serves as a publishing platform for scientific and technological approaches to "green"—i.e. environmentally friendly and sustainable—technologies. While a focus lies on energy and power supply, it also covers "green" solutions in industrial engineering and engineering design. Green Energy and Technology addresses researchers, advanced students, technical consultants as well as decision makers in industries and politics. Hence, the level of presentation spans from instructional to highly technical.

Indexed in Scopus. **Indexed in Ei Compendex**.

More information about this series at https://link.springer.com/bookseries/8059

Tariq Muneer · Eulalia Jadraque Gago ·
Saioa Etxebarria Berrizbeitia

The Coming of Age of Solar and Wind Power

Springer

Tariq Muneer
School of Engineering and Built Environment
Edinburgh Napier University
Edinburgh, Midlothian, UK

Eulalia Jadraque Gago
Engineering Construction and Project Management
School of Civil Engineering
University of Granada
Granada, Spain

Saioa Etxebarria Berrizbeitia
Faculty of Mechanical Engineering
University of the Basque Country
Vitoria-Gasteiz, Spain

ISSN 1865-3529 ISSN 1865-3537 (electronic)
Green Energy and Technology
ISBN 978-3-030-92009-8 ISBN 978-3-030-92010-4 (eBook)
https://doi.org/10.1007/978-3-030-92010-4

This Springer imprint is published by the registered company Springer Nature Switzerland AG
The registered company address is: Gewerbestrasse 11, 6330 Cham, Switzerland

Acknowledgements

T. Muneer would like to thank a number of people who have encouraged him to complete this project as they see a lot of benefits coming out of it, particularly for many countries where renewable energy will bring about a welcome change. Some of those names that come to mind are (in alphabetical order): Z. Ahsan and G. M. Khan of Malakpet Group; S. Farhatullah, A. Hussain, A. Hyder, M. W. Khan, M. H. Khursheed, F. Mirza, S. A. Samad, A. S. Taiyebi of Osmania University College of Engineering; and members of Edinburgh Breakfast Club—M. T. H. Esmail, H. Gazali, J. Mohammed, M. A. Mohammed, and K. Salamat.

Special thanks to Zafar Akber and Gulam Mohammed for their constant encouragement to complete this project. Their kind words have helped the author throughout the pandemic and its demoralising effects.

Contents

List of Figures

List of Tables

List of Textboxes

Chapter 1
Introduction

1.1 Introduction

Man is a weak animal and yet he dominates all animal kingdoms. The reason is that man can control forces that far exceed his muscle power. There lies man's quest to harness energy from sources that are external to his anatomy.

The earliest sources of energy that humans exploited were wind and sun. The history of wind energy may be traced up to 2800BC when the Egyptians used it to sail their ships. Textbox 1.1 charts the relevant historical development. Likewise, solar energy use has been recorded within the historical developments of the Greek civilisation. Archimedes, a Greek physicist allegedly used highly polished shields to focus the sunlight to burn down enemy ships. Later on, in the third century BC, the Greeks and Romans were reported to harness solar power with mirrors to light their torches. Chinese civilisation documented the use of mirrors for the same purpose later on in the year 20 AD.

Textbox 1.1 Historical development of wind resource exploitation

Wind is one of the oldest sources of power. Shown below is a synopsis of the historical developments against the quoted year.
2800 BC: Egyptians use sails to power their ships
700 AD: Persians started building vertical-shaft windmills for grain grinding
800–1400: Other Middle-Eastern civilisations passed the technology to Europeans
1850: American Farm Windmill developed, copied in Argentine and Australia
Early 1900s: Development of Dutch wind-powered electricity
1930s: 1-kW American wind-chargers were marketed
1940s: Smith-Putnam 175-ft, 1.25 MW twin-blade machine built in Vermont

T. Muneer et al., *The Coming of Age of Solar and Wind Power*, Green Energy and Technology, https://doi.org/10.1007/978-3-030-92010-4_1

1950–70: French built 100–300 kW units
2000: First 1 MW UK offshore machine installed at Blyth
2001–05: 2 MW machine became popular in Britain
2003: 3.6 MW machines commissioned at Arklow Bank wind farm
2006–08: 4–5 MW machines installed in UK and German waters
2018–19: Development of Siemens-Gamesa 10 MW & GE Renewable Energy 12 MW Haliade-X machines
2020: Development of Siemens-Gamesa 14 MW machines.

Another early passive use of solar energy was to admit sunlight via windows or glazing in walls and roofs to heat baths. This application can still be seen in some of the Turkish baths in Istanbul and other Turkish towns.

The invention of modern solar cells may be traced back to the work of Chapin, Fuller, and Pearson who created the first silicon photovoltaic cell at Bell Labs in 1954.

It is ironical that energy conversion technological developments have gone full circle, i.e. renewable energy use was superseded by coal, oil, gas, nuclear, then back to solar, wind, and wave power.

Presently employed energy systems will be unable to cope with future energy requirements—fossil fuel reserves are depleting, and climate change has become a serious issue. Figure 1.1 presents an updated profile of the greenhouse gases concentration in the atmosphere [1]. In the year 2019, the global average atmospheric carbon dioxide concentration was 409.8 ppm and these levels are higher than at any point in the past 800,000 years. In fact, the last time the carbon dioxide concentration was this high was more than 3 million years ago when the temperature was 2–3 °C higher than pre-industrial era and the sea level was 15–25 m higher than today [2].

Fossil fuel and nuclear energy production and consumption are closely linked to environmental degradation that threatens human health and quality of life and

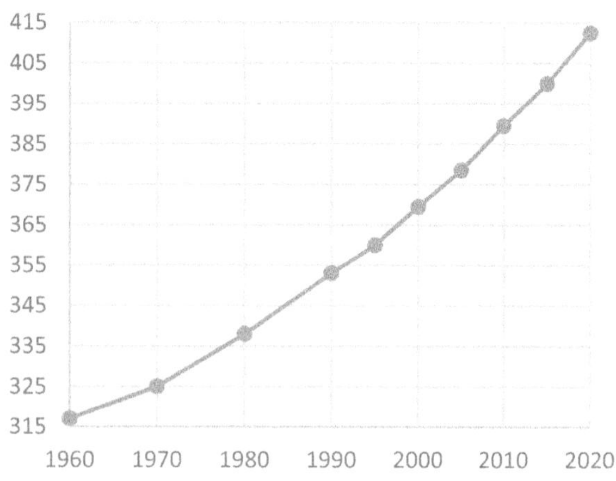

Fig. 1.1 Mean atmospheric carbon dioxide concentration at Mauna Loa Observatory. *Source* Ref. [1]

affects ecological balance and biological diversity. Although available only in a diffuse quantity, renewable energy is abundant, inexhaustible, and widely available. These resources have the capacity to meet the present and future energy demands of the world as indicated in Fig. 1.2 [3]. The cost of energy generated from these renewable resources is significantly falling as shown in Figs. 1.3 and 1.4 [4]. On the other hand, the cost of fossil fuel produced energy is in an increasing mode. Over the last three decades, solar and wind energy systems have experienced rapid growth. This is being supported by several factors such as declining capital cost; declining cost of electricity generated and continued improvement in performance characteristics of these systems. Figures 1.5 [5] and 1.6 [6] respectively present the growth of solar photovoltaic (PV) and wind energy installations in UK alone. This pattern is being duplicated the world over.

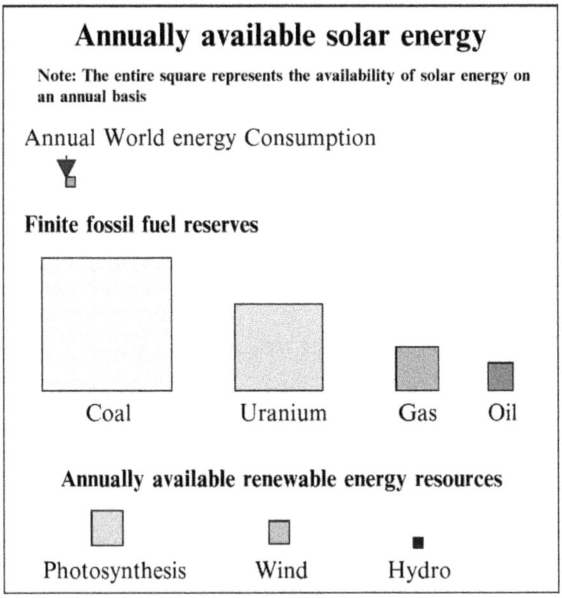

Fig. 1.2 Global energy resources and demands. *Source* Ref. [3]

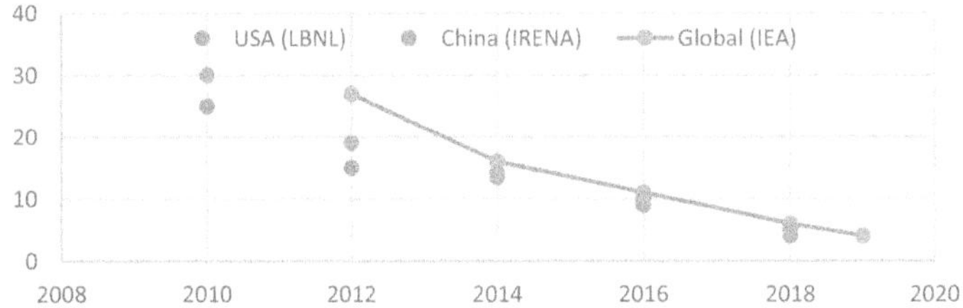

Fig. 1.3 Utility scale PV electricity price by year, US cents per kWh. *Source* Ref. [4]

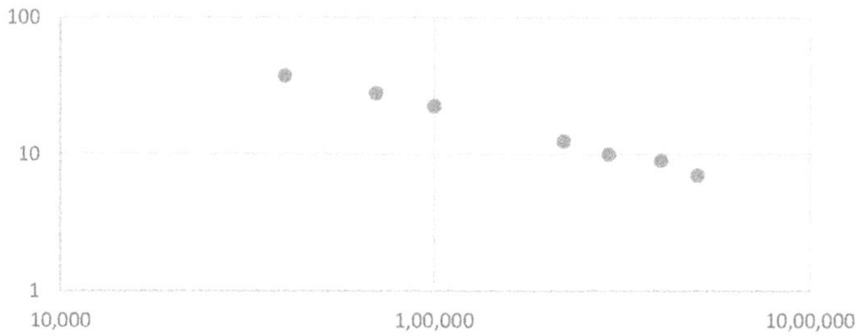

Fig. 1.4 Utility scale PV electricity price as a function of installation capital cost (log–log plot). X-axis: cumulative installed capacity, MW, Y-axis: price, US cents per kWh. Ref. [4]

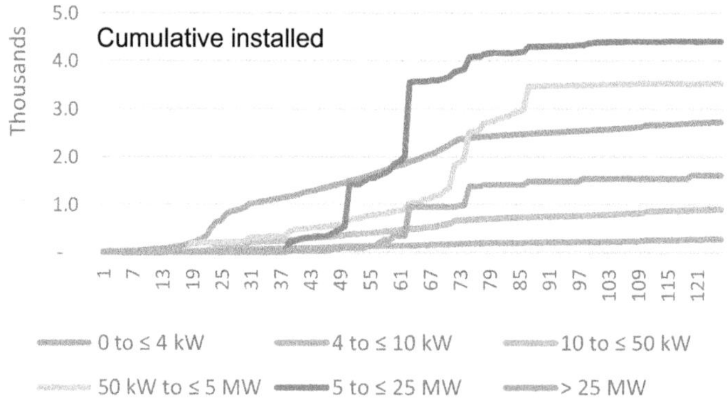

Fig. 1.5 PV installed capacity in UK between January 2010 and June 2020. X-axis: data series, Y-axis: thousand MW. Ref. [5]

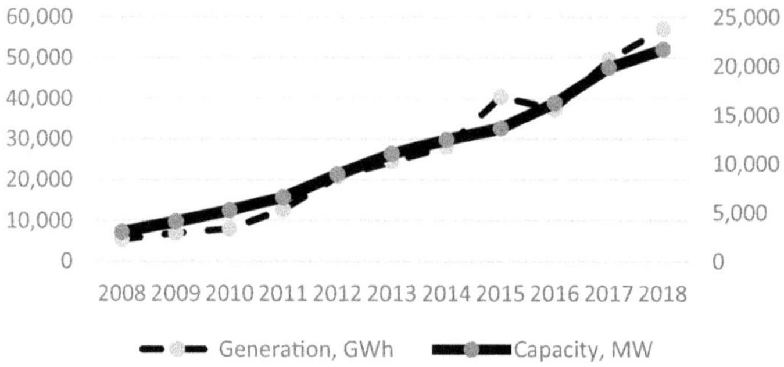

Fig. 1.6 Yearly progression of wind turbine installed capacity (MW, right axis) and generated energy (GWh, left axis) [6]

The cost of electricity from offshore wind projects completed during 2012–14 was UK pence 13.1/kWh compared to a wholesale price of UK pence 4–5/kWh. In 2017 the *Financial Times* reported that new offshore wind costs had fallen by nearly a third over four years, to an average of 9.7 UK pence/kWh, meeting the government's target of 10 UK pence/kWh four years early. Later in 2017, two offshore wind farm bids were made at a cost of 9.7 UK pence/kWh for construction by 2022–23. It is expected that a further 6GW of clean energy is to be added to the grid by year 2025 at around 4.7 UK pence/kWh at 2019 prices. These figures are lower than current generation costs for fossil fuels.

This book reviews the status of solar and wind energy in four regions/countries of the world that have managed to make rapid and exemplary progress in the installation of the latter renewable energy systems. Furthermore, the rapid technological advances are reviewed and mathematically rigorous procedures have presented that deal with the design and performance assessment of these energy systems.

This chapter shall present the need for a rapid switchover from fossil fuel to renewable energy systems and the associated issues.

1.1.1 Human Civilisation and Energy Use

Energy is one of the most basic of human needs. The accomplishments of civilisation have largely been achieved through the increasingly efficient and extensive harnessing of various forms of energy to extend human capabilities.

Providing adequate and affordable energy is essential for eradicating poverty, improving human welfare, and raising living standards worldwide. With the exception of humans, every organism's total energy demand is its supply of energy in the form of food derived directly or indirectly from sun's energy. For humans, the energy requirements are not just for heating, cooling, transport, and manufacture of goods but also those related to agriculture. However, the ingenuity of humans is such that throughout history there has been an exponential increase in the carrying capacity of land, particularly during the past few decades. Table 1.1 presents the remarkable progression of the agricultural land carrying capacity [7]. What is now needed is for mankind to achieve a 'sustainable energy revolution' that matches the dynamism of its agricultural counterpart. Note that over the last 10,000 years agricultural production has increased in intensity by a factor of a million. However, owing to the diffuse nature of renewable energy its generation will have to decrease in intensity. Let us examine the relative energy intensity of non-renewable and renewable energy sources from a required land area perspective. Two examples are cited here.

Operated by Drax Group, the 3.9 GW Drax co-firing power station in North Yorkshire, England occupies an area of 7.5 ha and uses biomass—(2.6 GW) and coal-fired plants (1.3 GW). The plant generates around 18 TWh of electricity per annum. This translates to a capacity factor of 53% which is somewhat lower than an average figure of 80% for most fossil-fuelled plants. The power and energy-generating densities are, therefore, respectively, 52 kW/m^2 and 240 MWh/m^2.

Table 1.1 Progression of agricultural land carrying capacity

Mode of agriculture	Number of humans supported/km^2
Hunter-gatherer	1
Domesticated animals	2
Shifting cultivation, S.E. Asia	10
Medieval agriculture (1200 CE, UK)	20
Intense tropical peasant agriculture (New Guinea)	50
Modern agriculture, UK	150
Modern agriculture, USA	500
Modern agriculture, Japan	2500
Phytofarming	1,000,000

The Drax Group now plans to switch coal-operated units to combined-cycle gas-turbine technology, with 200 MW of battery storage as well [8].

The Fukushima Daiichi Nuclear Power Plant is a disabled nuclear power plant located on a 3.5-km^2 land in Fukushima Prefecture of Japan. The plant suffered major damage from the magnitude 9 earthquake and tsunami that hit Japan on March 11, 2011. The chain of events caused radiation leaks and permanently damaged several reactors, making them impossible to restart. The lead author of this monograph visited the plant in year 1995 and was suitably impressed by the cleanliness of the plant and the use of waste heat to promote fish farms in warm waters, the growth rate of marine life being double of what nature could deliver.

The plant consisted of six boiling water reactors that drove electrical generators with a combined power of 4.7 GWe, making it one of the 15 largest nuclear power stations in the world. The capacity factor for nuclear power plants is quite high, reaching a value of 90% for well-maintained plants. Figure 1.7 presents information on capacity factors for nuclear power plants. Using an average capacity factor of 70% we get the power and energy-generating densities for this plant as 1.3 kW/m^2 and 8.2 MWh/m^2, significantly lower than those quoted above for a coal-fired plant [9]. That may be due to additional precautionary equipment and the storage space needed for a nuclear facility.

The Royal Academy of Engineering report [9] has reported power and energy density figures of 0.5 kW/m^2 and 3MWh/m^2 for nuclear power plants. The report also refers to the corresponding power and energy density figures of 3 W/m^2 and 9.2kWh/m^2 for wind energy generation. Thus, compared to nuclear, the power and energy densities for wind are almost two orders of magnitude lower.

Note that the capacity factor for offshore and onshore wind turbines for good locations are between 40–50% and 30–40% respectively. Indeed for the northern British islands, that figure approaches the ideal limit of 59%.

The average lifetime capacity factor at 28 operating UK offshore wind farms is 34.5%. There is a dependence of capacity factor on age, with older farms showing

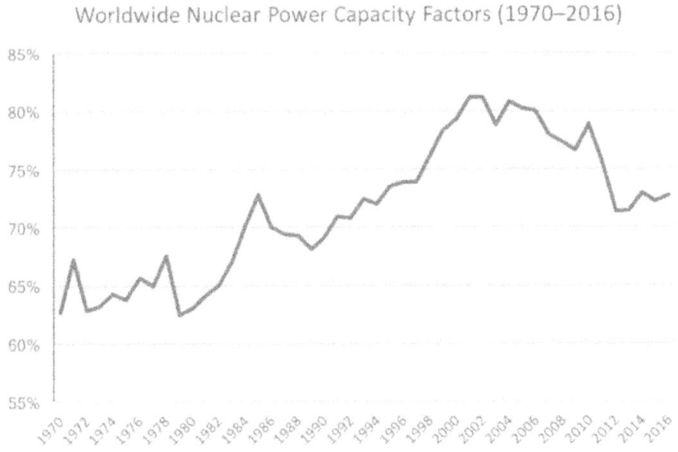

Fig. 1.7 Capacity factor for nuclear power plant

capacity factors of around 30% and younger ones' factors of around 40%. This is interpreted to be a result of increased turbine sizes, with taller modern turbines accessing higher wind speeds at higher elevations. There is no evidence for significant degradation of turbine performance with time. More recent UK offshore wind farms coming on line demonstrate a capacity factor of around 41% [10].

Likewise, data obtained from Danish wind farms that have been operational for over 16 years, with turbines rated at an average of 2.3 MW peak indicate a lifetime capacity factor of 44.3%. One of these offshore wind farms, Vindeby has achieved a service lifetime of 25 years [11].

Like wind, solar energy is also diffuse in nature. The corresponding capacity factors, like wind energy, strongly depend on latitude, varying between 10 and 25% as one moves from a latitude of 56 degrees to latitudes that lie within the equatorial belt.

To date, Scotland's largest (13 MWp) solar farm on Errol Estate in Perthshire (shown in Fig. 1.8) has been providing power for more than 3500 homes since May 2016. The farm occupies an area of 72 acres and this results in an energy density of 43 kWh/m^2-annum. The corresponding power density is 45 W/m^2. These are one order of magnitude higher than those for wind resources, the reason being the latitudinal and longitudinal spacing that is required between turbines on any given wind farm to avoid turbulence in the wake of individual machines. In this respect, one may refer to Fig. 1.9 which shows the recommended spacing between the machines [3, 12].

Figure 1.10 demonstrates the complementary nature of the solar and wind resource. On any given latitude the sum of the two resources must add up to a near constant value, with locations receding from the equatorial belt being richer in the latter resource and vice-versa.

Fig. 1.8 Errol solar PV farm in Perthshire, Scotland

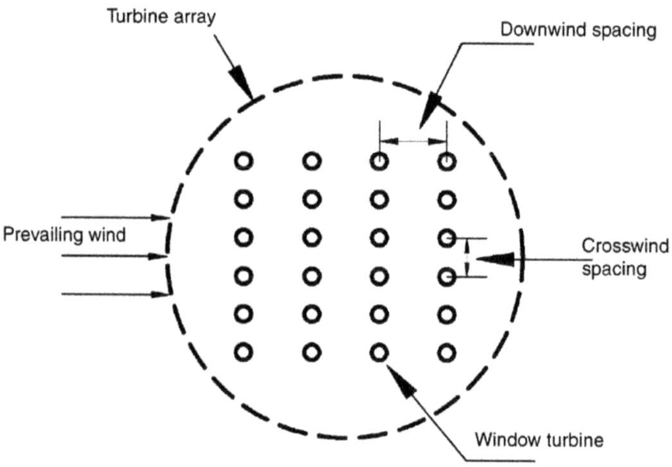

Fig. 1.9 Wind farm array schematic. The recommended crosswind and downward spacing for achieving less than 10% farm energy loss are respectively, 5 and 10 rotor diameters between the turbines. *Source* Asif and Muneer [3]

Figure 1.11 shows an inherent link between prosperity and energy use. Given the past record of developed countries in their profligate use of energy, developing nations tend to mimic their energy consumption pattern to match those of the developed nations. What Fig. 1.11 does is help us increase our understanding of the expected global energy consumption increase as nations across the world target their quest for increased prosperity. If increased life expectancy is accepted as an important index of well-being, then Fig. 1.11 shows that well-being is achievable at much lower levels of energy use index. Any further energy use may then be regarded as either sheer luxury or energy wastage. An example of this type of wastage could be single occupancy, large automobiles being driven aimlessly.

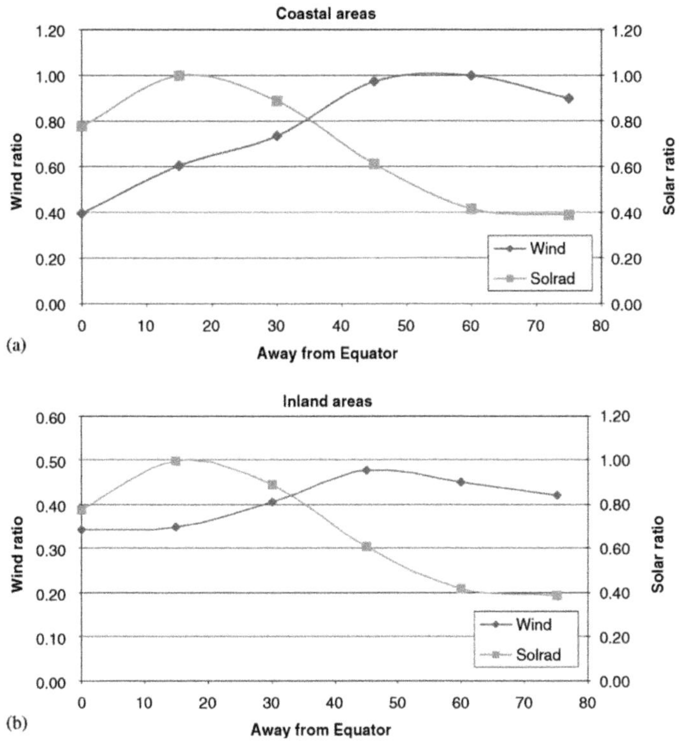

Fig. 1.10 The complementarity of solar and wind resource. *Source* Asif and Muneer [3]

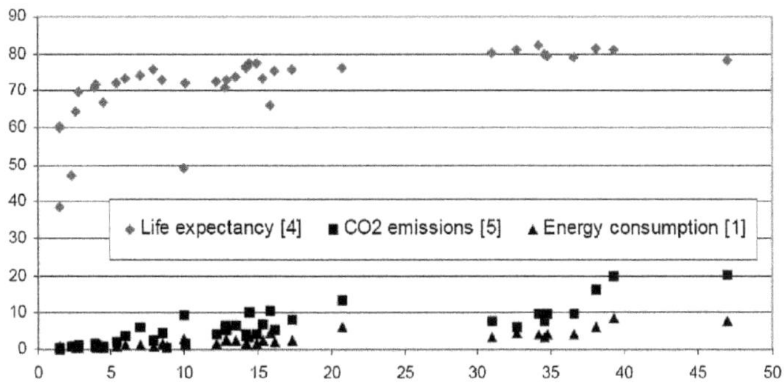

Fig. 1.11 Key performance indicators for selected nations plotted against GDP (thousand USD per capita, year = 2008). Life-expectancy in years, year = 2005; Carbon dioxide emissions, tonnes per capita, year = 2004; energy consumption, toe per capita, year = 2005

1.1.2 World Population

It has been estimated that the world population today is 1860-times the size of what it was 12 millennia ago when the world population was just around 4 million.

Table 1.2 shows the manner in which the above increase in population has occurred [13]. Between 10,000 BCE and 1700 CE, the population grew slowly at an annual rate of 0.04%. The world population was around 1 billion in the year 1800 and has increased seven-fold since then.

Around 108 billion people have ever lived on planet Earth which means that today's population makes up only 6.5% of the total number of people ever born.

For the long period from the appearance of modern Homo sapiens up to the starting point of Table 1.2 in 10,000 BCE, it is estimated that the total world population was often well under one million.

Man's infatuation with energy exploitation is not very old. As a matter of fact, up to the year 1769, coal was not used for mechanical power production. That was the year when James Watt patented his steam engine thus fuelling the industrial revolution. By the year 1800, the emissions from coal burning had risen to 0.05 Giga-tonnes of carbon dioxide per year. From then on the rate of Britain's emissions took off in a serious manner doubling by the year 1830 and then doubling again by 1850. The world carbon dioxide had exceeded 3.6 Giga-tonnes per year by year 1900 [13]. By the year 2019, those annual emissions had risen to 36.8 Giga-tonnes. This human lust for energy use may be related to one of the tales from Arabian Nights, depicted presently via the cartoon of Fig. 1.12.

For energy planners, there are seven factors that are of paramount importance. These may be enumerated as:

Table 1.2 Growth of human population

Years	Population (million)	Notes
10,000 BCE	4	
0	190	
1700	600	The average growth rate between 10,000 BCE and year 1700 was 0.04%
1800	990	During mid-fourteenth century the Black Death pandemic killed 200 million people in Europe
1900	1650	
1928	2000	
1960	3000	
1975	4000	
1987	5000	
1999	6000	
2011	7000	
2019	7700	

Fig. 1.12 The liberation of genie of the lamp. Once released, the genie cannot be compelled back in the lamp. This is akin to human addiction to energy use

- Present population.
- Projected population for the near and far-terms which will require population growth rate.
- Energy consumption per capita includes present consumption as well as the rising expectations of the developing world.
- The energy mix and its environmental impact.
- Land and sea area required for deployment of energy plants.
- Energy distribution systems which include grid design.
- Monetary capital—as well as total cost that additionally includes environmental, servicing, and maintenance issues.

Table 1.3 presents data on the global population growth [14]. The table shows that global population growth reached a peak in 1963 with an annual growth rate of 2.2%; but since then, world population growth has halved and the forecast is for it to drop further, reaching 0.75% in year 2030 and then further halving to 0.375%

Table 1.3 Annual world population growth rate

Year	Growth rate (%)
1963	2.20
1980	1.70
1990	1.55
2000	1.25
2010	1.13
2019	1.10
2030	0.75
2060	0.375

by 2060. The latter two time frames will be of significance to energy planners as respectively being near and medium terms. It has been estimated that by years 2030 and 2060 respectively, 8.6 and 9.3 billion humans will populate the planet.

Table 1.4 presents population, geographical area, and energy usage data for the selected countries/regions of the world [15–19].

For energy planners and policy makers those statistics are of importance.

1.1.3 Climate Change and Its Impact

This section is based on reference [20]. Climate hazards are natural events that occur in a cyclical manner. Hurricanes, droughts, wildfires, flooding, and storms accompanied by high winds have always been around. However, of late the scale of destruction seems to be intensifying.

For example, recent climate disasters that may be cited are Cyclones that hit New Orleans, Haiti, The Caribbean, and Bangladesh, heat waves in South Asia and Europe, and flooding in numerous cities across the world.

Climate Change has caused increases in air and water temperatures, raised sea levels, potent storms, and higher wind speeds. The evidence is overwhelming and the results devastating. To enumerate:

- The number of climate-related disasters has tripled in the last 30 years.
- Between 2006 and 2016, the rate of global sea-level rise was 2.5 times faster than it was for almost all of the twentieth century.
- More than 20 million people a year are forced away from their homes by climate change.
- The United Nations Environment Programme (UNEP) estimates that adapting to climate change and coping with damages will cost developing countries $140–300 billion per year by 2030.

Table 1.4 Population, geographical area and energy statistics for selected countries

	2017	EU28	USA	China	Japan	Total	World	Percent of world (%)
Population*	(Million people)	446	325	1386	127	2284	7510	30
Population growth	(%)	0.15	0.63	0.51	−0.16			
Area*	(Thousand km^2)	4239	9147	9388	365	23,139	127,355	18
GDP/capita+	(k$)	33.0	60.0	8.8	38.3			
GDP/growth rate	(%)	2.7	2.2	6.8	1.9			
CO_2 emissions++	(Metric tonne/capita)	6.3	15.9	6.7	8.9			
Energy consumption/capita	(kg oil equivalent)	3792	6839	2265	3590			
Energy consumption/capita	(kWh)	44,100	79,536	26,339	41,757			
Total energy consumption[a]	(TWh)	19,674	25,848	36,505	5294	87,322	156,708	56
Electricity consumption	(TWh)	784	782	3066	208	4839	10,698	45
Coal consumption	(TWh)	2724	3853	21,985	1394	29,957	43,245	69
Oil consumption	(TWh)	7554	10,490	7102	2184	27,331	53,579	51
Gas consumption	(TWh)	4657	7394	2404	1170	15,625	36,540	43
Nuclear consumption	(TWh)	2184	2229	652	77	5143	6944	74
Renewable energy consumption	(TWh)	1772	1099	1296	261	4428	5701	78

*https://data.worldbank.org/

**https://data.worldbank.org/indicator/AG.LND.TOTL.K2?locations=1W

+https://data.worldbank.org/indicator/NY.GDP.PCAP.CD?locations=US

++https://www.iea.org/data-and-statistics/?country=CHINAREG&fuel=CO2%20emissions&indicator=CO2%20emissions%20by%20energy%20source

[a]https://www.bp.com/content/dam/bp/business-sites/en/global/corporate/pdfs/energy-economics/statistical-review/bp-stats-review-2019-full-report.pdf

An average value of the latter loss, just for the developing countries, may be taken as $220 billion and this economic loss is to be borne in mind when the cost of energy generation from fossil fuels is accounted.

Specific examples of what is being understood as climate change induced disasters may be quoted here.

African cyclone: In March 2019, Cyclone Idai took the lives of more than 1000 people across Zimbabwe, Malawi, and Mozambique in Southern Africa. Lethal landslides took homes and destroyed land, crops, and infrastructure.

Australian wildfires: The start of 2020 found Australia in the midst of its worst-ever bushfire season that followed from its hottest year on record. The fires burned through more than 10 million hectares and left millions of people affected by smoke. More than a billion native animals were killed.

East African drought: Higher sea temperatures, linked to climate change, have doubled the likelihood of drought in the Horn of Africa region. Severe droughts in 2011, 2017, and 2019 have repeatedly wiped out crops and livestock. Droughts have left 15 million people in Ethiopia, Kenya, and Somalia in need of aid, yet the aid effort is only 35 percent funded. People have been left without the means to put food on their table, and have been forced from their homes. Millions of people are facing acute food and water shortages.

South Asian floods: Over the last year deadly floods and landslides have forced 12 million people from their homes in India, Nepal, and Bangladesh. Just 2 years ago exceptionally heavy monsoon rains and intense flooding destroyed, killed, and devastated lives in the same countries. In some places the flooding was the worst for nearly 30 years, a third of Bangladesh was underwater. While some flooding is expected during monsoon season, scientists say the region's monsoon rains are being intensified by rising sea surface temperatures in South Asia.

Dry Corridor in Central America: An El Niño period, supercharged by the climate crisis, has taken Central America's Dry Corridor into its 6th year of drought. Guatemala, Honduras, El Salvador, and Nicaragua are seeing their typical three-month dry seasons extended to six months or more. Most crops have failed, leaving 3.5 million people, many of whom rely on farming for both food and livelihood, in need of humanitarian assistance, and 2.5 million people food insecure.

Up to the year 2050, global GHG emissions are expected to grow by 50% chiefly due to energy demand fuelled by economic growth in developing countries. Unless urgent action is taken concentration of GHG in Earth's atmosphere is likely to reach 685 ppm. That is over the concentration level of 450 ppm required to have at least a 50% chance of stabilising the climate at 2 °C. The latter was the goal set in 2010 at the UN Framework Convention on Climate Change conference in Cancun, Mexico [21].

Greta Thunberg began a lone protest outside Sweden's parliament in October 2018 when she was 15, saying would refuse to attend school on Fridays until the government tackled the growing climate and ecological crisis. In the 12 months since, she has become one of the world's most talked-about people, having been invited to speak at the UN general assembly and inspiring global protests attracting hundreds of thousands of young campaigners.

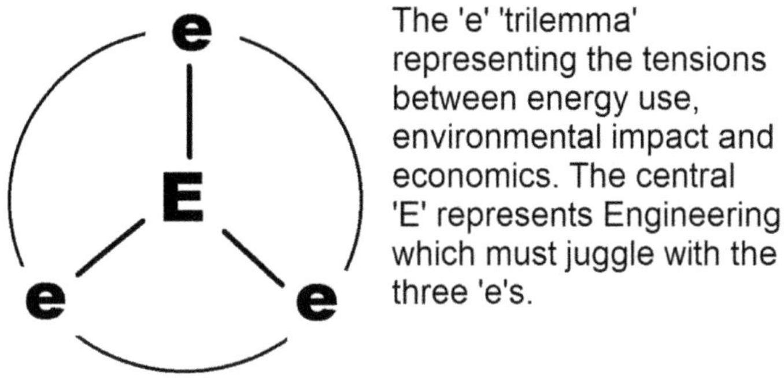

Fig. 1.13 The three 'e' trilemma

For any good argument to be put forward for achieving energy sustainability it is important to critically examine the hard facts related to the present fossil-fuelled and upcoming renewable energy generation. The costs with respect to energy, economics (monetary), and environment need to be audited. This monograph is an attempt to provide some of the relevant arguments, data, and procedures.

1.1.4 Electricity Use, Its Efficiency, and Related Issues

Figure 1.13 presents the 'e trillema'. Engineering has been responsible for the industrial and energy revolution and thus has been tasked with juggling energy, environment, and economics.

In this section we shall look closely at the increasing use of electrical energy both within the building and transport sector driven mainly by heat pumps to heat- and refrigeration for cooling of buildings, and the increasing adaption of electric vehicles. The ground is thus being prepared for electricity as a replacement for coal, oil, and gas. If the increasing amount of electrical energy that will thus be required in the coming decades was sourced from fossil fuel burning that will only bring forward the tipping point of climate change disaster. The only solution will be to bring in an increasing capacity of renewable and/or nuclear energy. The cost, safety, and other environmental impacts of nuclear power, however, need to be discussed at length though and this monograph can only scratch the surface of those issues. A large amount of literature is available on the latter topics towards which the reader may be directed [22].

It has long been debated that nuclear power plants produce next to nil carbon emissions. However, a closer look at the latter source of electricity shows a different picture. Due to the relevant safety issues, the construction of those power plants contains a very significant amount of embodied energy. Furthermore, at best those plants are 40–45% efficient and hence the bulk of the extracted energy is waste heat which is dumped in lakes, lagoons, or sea. Figure 1.14 shows solubility data for

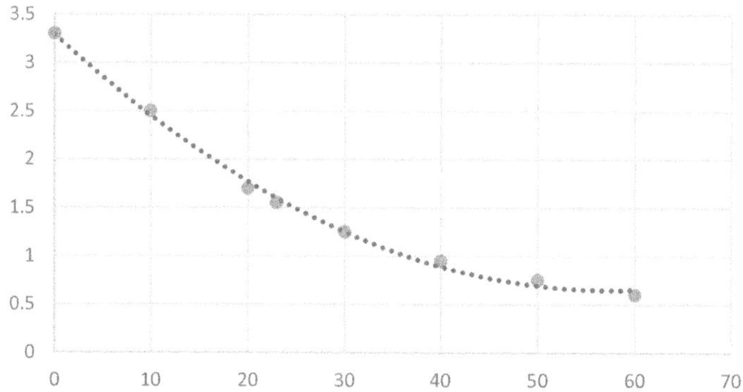

Fig. 1.14 Solubility of carbon dioxide in water

carbon dioxide in water. We observe that between10 and 25 °C there is a sharp decline in solubility and hence carbon dioxide will be ejected from those water bodies where waster heat is being discarded. Note that the Earth's seas hold up to 50 atmospheres of carbon dioxide [23]. In this respect, the reader is referred to one of the present author's press notes [24, 25].

In view of the climate change related arguments related to fossil fuel burning, the sharp decline of the monetary cost of renewable energy and the energy density offered by wind turbines and solar PV, respectively upwards of 35 to 15%, as opposed to biomass which only offers efficiency of less than 0.3% the switch to wind/solar, is imminent.

1.1.4.1 Carbon Capture Technology

Carbon capture and storage technology are being discussed as a means to allow continued use of fossil fuel burning in electrical power plants. We find a number of procedures for evaluating the cost of carbon dioxide capture, storage, and disposal. These may be enumerated as the cost calculation of carbon dioxide (a) avoidance, and (b) capture. The cost of avoidance is defined as the average cost of avoiding a unit emission to the atmosphere while producing a kWh of electricity.

The cost of electricity of the plant is the levelised cost of electricity produced (US or Euro cents/kWh). The cost of electricity production with the retrofitted capture, storage, and disposal of carbon dioxide facility is dependent on project life, salvage value of plant, construction period, capacity factor, interest rate, fuel costs, and cost associated with the chemistry of the process [26].

Although carbon dioxide capture technologies are commercially available for use on coal-fired power plants, they have substantial capital and operating costs. A pulverised coal plant with a commercially available amine-based carbon dioxide capture system that is designed to remove 90% carbon dioxide requires 27% of plant capacity to capture carbon dioxide and compress it for sequestration. That is indeed a high proportion of parasitic loss of energy. Capital costs represent 42% of the total plant cost. The incremental cost of production associated with carbon dioxide removal is 4 US cents/kWh, and the cost per ton of carbon dioxide removed is $47 [27].

The Carbon Capture & Storage (CCS) technology also claims that it can capture up to 90% of the carbon emissions from fossil fuel based electricity plants.

The Carbon Capture & Storage Association estimated that the relevant cost to be $69–$103 per tonne of carbon dioxide abated [28].

Yincheng et al. [29] have carried out a survey of several technologies of carbon dioxide sequestration that include absorption, adsorption, cryogenic methods, and membrane separation, and biological fixation. They have alluded that the absorption process is one of the most common industrial technologies that shows good promise. However, the cost to capture carbon dioxide from power plants remains high. The estimates presented by them show that the energy penalty from using one of the leading absorption processes is 15–35%.

To summarise from above cited literature:

- A 27% plant capacity parasitic energy loss is needed to capture, remove 90% of carbon dioxide exhaust and compress it for sequestration.
- The relevant capital costs represent 42% of the total plant cost.
- The incremental cost associated with carbon dioxide capture, removal, and compression technology removal is 4 US cents/kWh.
- The energy penalty from one of the leading absorption processes is 15–35%.

The above cost estimates have to be incorporated within the electricity generation economics.

The issues related to wind/solar based energy resources that need to be addressed are: generation technologies, and transmission of electrical power and storage, the latter being of particular concern in view of their highly intermittent nature. As a matter of fact, fossil fuel power plants also needed storage but that was available in the form of stored fuel, coal-bunkers, oil or gas storage tanks. In view of terrorism, those storage facilities have however attracted attention, indeed those facilities are now being considered undesirable.

1.1.4.2 Energy Use and Energy Efficiency

Published in September 2019, The Oxford Institute for Energy Studies report on 'Glimpses of China's Energy Future' shows that the year 2020 total energy use in China is on target to be 39.5 PWh. Note that one PWh is a million, million kWh. That energy use is continuing at an annual rate of 2.4%. Likewise, the 'Energy, transport and environment statistics 2019 edition' captures the energy statistics data for the EU28. The energy demand within the EU28 states has stabilised with a net decrease in demand of 0.1% recorded between 2016 and 2017, the latest period when the energy statistics were gathered.

Table 1.5 Final energy use in two of the chosen economies (China* and EU**)

	Buildings	Transport	Industry	Other	Total	Population, million	kWh/capita
China	4.7	4.7	19.2	11.0	39.5	1386	28,530
EU28	3.2	3.6	2.9	2.0	11.6	446	26,076

*The Oxford Institute for Energy Studies report on 'Glimpses of China's Energy Future', September 2019

**Energy, transport and environment statistics 2019 edition', Luxembourg: Publications Office of the European Union, 2019

Table 1.5 presents a sectorial breakdown of the energy use for world's largest, developing economy (China) and a mature and developed economy (EU28) [30].

There are several important points of note that are enumerated below:

China, with its nearly triple population load, is presently consuming similar amount of energy as compared with EU28 for two of the four sectors, i.e. buildings and industry. Those figures are bound to increase for China as its standard of living increases further.

China seems to consume nearly six fold more energy for the other two sectors—industry and 'other' category as it is in a developing stage. Those quantities may level off as it attains a 'developed' status.

The present per capita energy use is nearly equal for China and the EU28. However, that figure is somewhat deceptive in view of an asymmetrical consumption recorded by the industry.

As shown in Table 1.5 once the economies of any given country are developed a stable energy consumption level is reached, with the two chief sectors being energy use in buildings and transport. Most appliances that are responsible for energy use in buildings have or are reaching an asymptotic level of efficiency. Examples that may be cited here are refrigeration and air conditioning for cooling, building heating plants, electric motors, and electric lighting. In this respect, Table 1.6 shows the lumi-

Table 1.6 Luminous efficacy of selected light sources

Technology	Luminous efficacy (lm/W)	Energy conversion efficiency (%)
Candle	0.01	
Oil lamp	0.03	
Filament tungsten lamp, 60 W	Between 10 and 18	2
High voltage halogen lamp, 120 W	18	12
Low voltage halogen lamp, 60 W	26	15
Fluorescent lamp, 54 W	82	24
LED lamp, 16 W	151	49

Table 1.7 Estimated levelised cost of electricity for new generation resources entering service in 2025. All cost figures are in 2019 US cents/kWh

Technology	Capacity factor (%)	Capital cost	O&M	Transmission cost	Total system LCOE
Ultra-supercritical coal	85	4.76	2.77	0.12	7.64
Combined cycle	87	0.84	2.85	0.12	3.81
Combustion turbine	30	1.62	4.70	0.35	6.66
Advanced nuclear	90	5.61	2.44	0.11	8.17
Geothermal	90	2.04	1.56	0.15	3.75
Biomass	83	3.99	5.37	0.13	9.48
Onshore wind	40	2.96	0.75	0.28	4.00
Offshore wind	44	9.10	2.87	0.27	12.23
Solar PV	29	2.61	0.60	0.36	3.57
Hydroelectric	59	3.73	1.36	0.19	5.28

Source US Energy Information Administration, Annual Energy Outlook, 2020

nous efficacy and efficiency of artificial lighting which has increased significantly in performance.

The most common definition of luminous efficacy, lumen per Watt (lm/W) of a light source is its luminous flux divided by the electrical power it consumes. The luminous efficiency of a light source is usually defined as its luminous efficacy divided by the maximum possible value of the efficacy. The ideal value of 683 lm/W is achieved for an ideal light source operating at a wavelength of 555 nm [31].

Table 1.7 displays the estimated levelised cost of electricity for new generation plants, the most economical being solar PV, followed closely by combined cycle and onshore wind [32, 33]. Note that both onshore and offshore wind generation costs are subject to capacity factors and the size of the turbine. The most recent Siemens, largest-to-date 14 MW offshore turbine that is being deployed in the North Sea is likely to offer higher capacity factor and lower generation cost. The latter subject shall be dealt with in the following chapter in greater detail.

References

1. Scripps Institution of Oceanography/NOAA Global Monitoring Laboratory, June 2020
2. www.climate.gov
3. Asif M, Muneer T (2007) Energy supply, its demands and security issues for developed and emerging economies. Renew Sustain Energy Rev 11:1388–1413
4. https://rameznaam.com/2020/05/14/solars-future-is-insanely-cheap-2020/
5. https://data.gov.uk/dataset/9238d05e-b9fe-4745-8380-f8af8dd149d1/solar-photovoltaics-deployment-statistucs
6. https://en.wikipedia.org/wiki/Wind_power_in_the_United_Kingdom

7. Cartledge B (1992) Monitoring the environment. Oxford University Press
8. drax.com accessed on Aug 20, 2020 at 1420 hours
9. Wind energy: implications of large-scale deployment on the GB electricity system, Apr 2014
10. http://euanmearns.com/uk-offshore-wind-capacity-factors-a-semi-statistical-analysis/
11. https://energynumbers.info/uk-offshore-wind-capacity-factors
12. Database of the Global Environment and the United Nations (ourworldindata.org)
13. MacKay DJC (2009) Sustainable energy without the hot air, UIT, Cambridge, England
14. US Census Bureau and Projections, United Nations Population Division (Medium variant, 2015 revision) (ourworldindata.org)
15. https://data.worldbank.org/
16. https://data.worldbank.org/indicator/AG.LND.TOTL.K2?locations=1W
17. https://data.worldbank.org/indicator/NY.GDP.PCAP.CD?locations=US
18. https://www.iea.org/data-and-statistics/?country=CHINAREG&fuel=CO2%20emissions&indicator=CO2%20emissions%20by%20energy%20source
19. https://www.bp.com/content/dam/bp/business-sites/en/global/corporate/pdfs/energy-economics/statistical-review/bp-stats-review-2019-full-report.pdf
20. https://www.oxfam.org/en/5-natural-disasters-beg-climate-action
21. https://www.eia.gov/outlooks/aeo/pdf/electricity_generation.pdf
22. https://www.sciencedirect.com/topics/engineering/cost-of-electricity-production
23. www.engineeringtoolbox.com
24. Blow for N-plants on global warming (1991) The Scotsman, p 3, 3 June
25. Researchers at Napier University ... (1997) The Herald, p 26, 18 Feb
26. https://www.modernpowersystems.com/features/featurecan-ammonia-scrubbing-cut-the-cost-of-co2-capture-/
27. Report from Electric Power Research Institute and Parsons Infrastructure & Technology Group
28. https://www.power-technology.com/features/carbon-capture-cost/
29. Yincheng G, Zhenqi N, Wenyi L (2011) Comparison of removal efficiencies of carbon dioxide between aqueous ammonia and NaOH solution in a fine spray column. Energy Proc 4:512–518
30. The Oxford Institute for energy studies report on 'Glimpses of China's energy future', Sept 2019
31. https://www.rp-photonics.com/luminous_efficacy_and_efficiency.html
32. Energy, transport and environment statistics: 2019 edition (2019) Publications Office of the European Union, Luxembourg
33. Saravanan D (2018) Design of a residential solar PV system for an Indian dwelling, Edinburgh Napier University, Master of Science Thesis

Chapter 2
Solar and Wind Energy Technologies

In its December 2019 report 'Statista' noted that the global installed electricity capacity in year 2017 was 4.15 TW for fossil fuels and 1.01 TW for installed renewable sources [1]. Electricity generation from fossil fuels has consistently decreased over the last few years, whereas renewable electricity generation has, in turn, increased to meet demand. The cost of renewable energy sources has decreased over time, making them increasingly cost effective and deployable.

Global consumption of electricity has increased over the last decades from 7323 TWh in 1980 to 22.3 TWh in 2017. That is more than 8% increase per annum compared to other forms of energy use that have shown a 2% increase. Renewable sources now account for almost 27% of global electricity generation [2]. One point of note: in the coming years due to the introduction of electric vehicles in increasing large numbers the electricity demand will be much higher than 8% per annum increase cited above. This item shall be discussed at length in a later section of this chapter.

Europe has recently sailed past one of the major milestones of any major energy transformation program. London-based Ember has published a report outlining that renewables now account for a greater share of the energy mix than fossil fuels, and are now firmly the EU's main source of electricity. In the first half of 2020, just 34% of Europe's power came from coal, gas, and oil, versus the 40% that came directly from solar, wind, and hydro sources. Use of renewable energy also went up by over a tenth during that same period, and carbon dioxide emissions fell by nearly a quarter [3].

The Digest of UK Energy Statistics (DUKES) provides details on British energy use [4]. The DUKES revealed that in 2019 the UK generated 37.1% of all energy from renewable sources. The combined domestic output of wind, solar, and hydropower is documented as being up by 11% when compared to the previous year, and bioenergy output is also up by nearly 2%.

On the environment front, energy-related carbon dioxide emissions were at an all-time high of 33.1 Gt in 2018. While emissions from all fossil fuels increased, the electricity sector now contributes to two-thirds of emissions growth. Coal use in

T. Muneer et al., *The Coming of Age of Solar and Wind Power*, Green Energy and Technology, https://doi.org/10.1007/978-3-030-92010-4_2

electricity generation alone surpassed 10 Gt carbon dioxide emissions mark, mostly driven by the economies of China, India, and the United States while emissions have started to decline in Germany, Japan, Mexico, France, and the United Kingdom [5].

2.1 Wind Energy

The global wind power capacity increased by 19% in 2019, with the addition of 60 GW which included around 54 GW onshore and 6 GW offshore turbine capacities. Offshore wind energy now plays an important role in the global market, accounting for 10% additions in 2019. The total wind power capacity now stands at a cumulative figure of 651 GW overall with 621 GW onshore and the rest offshore [6].

In the year 2020, the following were the top 10 countries of the world with the largest share of wind energy installations. The list has been arranged in descending order of installed capacity [7].

1. China
   China has capacity of 221 GW with over a third of global capacity. It has the largest stock of onshore wind farms with a capacity of nearly 8 GW.
2. United States of America
   The USA has 96.4 GW of installed capacity. The country has six of the 10 largest onshore wind farms.
3. Germany
   Germany has 59.3 GW capacity and has the highest capacity in Europe. Its largest offshore wind farms have a capacity of 582 MW.
4. India
   India has the second-highest wind capacity in Asia, with capacity of 35 GW. The country has the third- and fourth-largest onshore wind farms in the world, i.e. the 1.5 GW Muppandal wind farm in Tamil Nadu and the 1.1 GW Jaisalmer Wind Park in Rajasthan.
5. Spain
   With its 23 GW capacity, Spain provides 18% of its electricity supply through wind energy. However, in the recent past, the growth seems to have stalled.
6. United Kingdom
   The UK has a total capacity of just over 20.7 GW. It has six of the 10 highest-capacity offshore wind projects in the world.

   In October 2020 the UK government pledged £160 m to upgrade ports and factories for building turbines, setting a target for wind to produce enough electricity to power the equivalent of every UK home by 2030. The number of turbines will increase from the 2020 stock of 1800 to 7000 and capacity will increase from 10 to 40 GW. The UK aims to become the world leader in low-cost clean power generation.

7. France
   France has capacity of 15.3 GW. There are reports of France gradually moving away from nuclear power which has in the past provided 75% of its energy needs.
8. Brazil
   Brazil has the largest wind capacity in South America with its 14.5 GW installations and seems to increase rapidly in recent years. Wind power rates fourth in Brazil's energy mix having a share of 8%.
9. Canada
   With a total of 299 wind farms having 6596 turbines, Canada's capacity stands at 12.8 GW.
10. Italy
   Italy reached 10 GW in wind energy capacity in 2018. Most of the installations are concentrated in the south and on its islands.

Refer to Fig. 2.1. It is proposed that an ultra-large, wind farm be developed off the Aberdeen shore. There is a substantial wind farm related activity in that area as Aberdeen has been the hub of the UK oil and gas development activity. The dark area shown in Fig. 2.1 could host such a development which would cover an area of 417 km^2 in the North Sea. The coordinates of that area are: 57° N, 1° W; 57° N, 1.25° W; 57.25° N, 1° W and 57.25° N, 1.25° W.

VGB PowerTech of Essen, Germany has prepared a detailed report on the 'Development of the global and European electricity demand' for the years 2020 and 2040. They have shown that the worldwide and EU electricity demands for the year 2040 will respectively be 35,000- and 3500 TWh [8].

Using one of the best global wind resources the wind farm shown in Fig. 2.1 would be able to generate 845 TWh/annum, of electricity using 14 MW turbines. Thus, the proposed wind farm will be able to provide one-quarter of the electricity needs of the entire EU for the year 2040.

Note that energy modelling details are provided in a later section.

In year 2020, Siemens-Gamesa launched their biggest offshore wind turbine that is rated at 14 MW. With a respective rotor diameter and blade length of 222- and 108 m it has a swept area of 39,000 m^2. The machine has variable speed and its power is pitch-regulated [9]. A mathematical model for the power curve of this machine is given in Eq. (2.1.1),

$$\begin{array}{l} \text{For } WS < 3.81,\ P = 0 \\ \text{For } 3.81 < WS < 12.5,\ P = -0.0081 * WS^4 \\ \quad +0.2354 * WS^3 - 2.2901 * WS^2 \\ \quad +10.136 * WS - 16.684 \\ \text{For } 12.5 < WS < 25,\ P = 14 \\ \text{For } WS > 25,\ P = 0 \end{array} \tag{2.1.1}$$

In Chap. 1 it was shown that nuclear power plants have capacity factors approaching a 90% mark. Figure 2.2 presents data from UK and Danish offshore

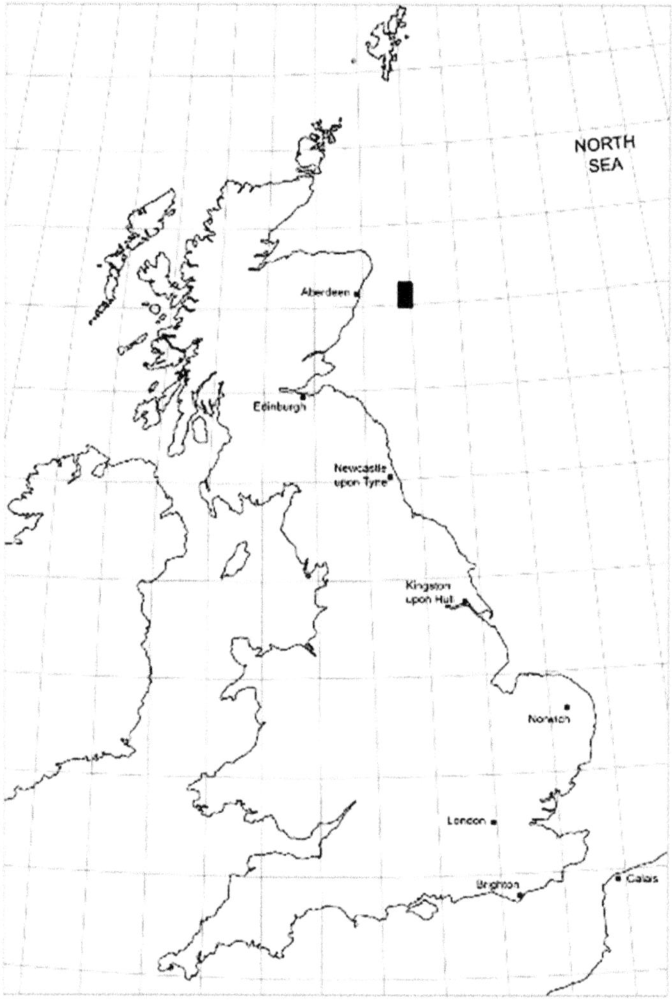

Fig. 2.1 Ultra-large wind farm

Fig. 2.2 Capacity factor—time of operation data for UK and Denmark offshore wind turbines

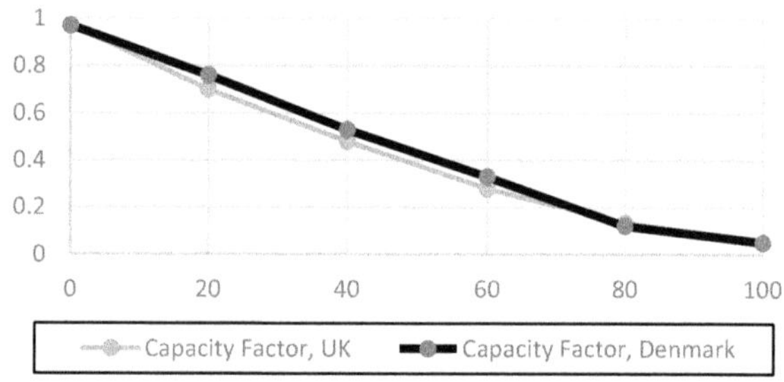

wind farms [10]. Table 2.1 and 2.2 presents information related to installed capacity and electricity generation for latterly mentioned farms. As a result of the accelerated deployment of wind farms the Carbon intensity of electricity is rapidly dropping (see Table 2.3).

The capacity factors for wind farms are obviously much lower than those that are achievable by nuclear- or fossil fuel power plants. Capacity factors for solar PV farms are even lower. To overcome the barrier of lower capacity factors, researchers from University of Delaware and Delaware Technical College [11] propose designing renewable energy farms that offshoot the energy requirements by a large amount. Under such a scheme wind and solar wind farms would be deployed to have generation capacities that far exceed the requirements. That would ensure the stability of energy supply in times of low resource, i.e. calm days experienced by wind farms and cloudy periods for solar PV fixtures.

The synopsis of the above cited research is:
A certain amount of wind and solar power plant capacity on a utility grid will meet 100% of electricity demand one day out of the entire year. That would appear to be the most cost-effective option. However, in view of the higher storage costs compared to the rapidly declining generation cost from wind/solar energy, a larger amount of wind and solar power plant capacity on a utility grid will meet all of the electricity demand for more days of the year. The larger the template capacity of wind/solar plants the longer period when the demand side becomes autonomous. Obviously, it costs more money to build ever larger wind and solar power plants. However, if the backup energy generation or storage systems cost more than the over-designed part of renewable energy system then the latter solution is more cost effective. The above conclusions may, however, change with time as:

- The electric battery storage costs are also declining (see Fig. 2.3).
- The number of electric vehicles is increasing very rapidly, with each new model coming out with higher battery capacity (Table 2.4).

Table 2.5 presents data for the cost of electricity in selected countries that are passed on to the consumer. It is highly probable that the costs shown for countries included at the bottom of the referred table include a heavy subsidy. The cost for countries at the top end, in all likelihood, represent the real cost for generation and supply. With that in mind, if we examine the electricity generation cost for solar PV and wind that was presented in Chap. 1, we begin to see the reason for the rapid development of these two sources of renewable energy. In this respect refer to the introductory section of Chap. 1 where the year 2025 offshore wind energy cost of 4.7 UK pence/kWh (US cents 6.6/kWh) was quoted. Furthermore, Table 1.7 indicated the year 2019 onshore and offshore per kWh costs of 4- and 12.23 US cents, respectively. Figure 1.3 provided a solar PV cost of 4 US cents/kWh.

Resource assessment and elementary modelling: wind energy

There are a large number of sources that are available to obtain data related to solar and wind energy resources. For precise information for a particular location, the local

Table 2.1 Generation and capacity factor data for UK wind farms

	Age, years	Life capacity factor	Installed capacity, MW	Generated electricity, GWh	Power/swept area, W/m^2
Aberdeen EOWDC	1.5	0.37	93	455	1.7
Barrow	13.3	0.36	90	3772	3.2
Beatrice	0.6	0.492	588	1501	2.2
Burbo bank	12.2	0.337	90	3252	3
Burbo bank 2	2.8	0.397	254	2439	2.5
Dudgeon	2.3	0.475	402	3770	3.5
Galloper	1.7	0.447	353	2416	1.4
Greater gabbard	6.4	0.406	504	11,606	1.4
Gunfleet sands	9.5	0.365	173	5313	3.9
Gwynt-y-Mor	4.5	0.344	576	7955	2.9
Humber gateway	4.6	0.432	219	3806	3.5
Hywind Scotland	2.2	0.538	30	317	1.1
Lincs	6.3	0.42	270	6276	2.8
London array	6.7	0.392	630	14,751	2
Ormonde	7.9	0.377	150	3908	5.7
Race bank	1.9	0.415	573	3995	3.8
Rampion	1.7	0.29	400	1776	1.6
Robin rigg	9.3	0.351	174	5025	3.4
Sheringham Shoal	6.3	0.399	317	7037	3.6
Thanet	9.3	0.33	300	8064	2.8
Walney phase 1	8.5	0.401	184	5515	2.6
Walney phase 2	7.5	0.456	184	5546	1.9
Walney 3 & 4	1.3	0.486	659	3898	2.1
Westermost Rough	4.6	0.47	210	3998	2.8
West of duddon sands	5.2	0.457	389	8073	2.7
Beatrice demo (closed)	8.3	0.189	10	123	
Blyth demo (closed)	12.8	0.087	4	37	
Blyth demo Array 2	1.5	0.42	42	207	

(continued)

Table 2.1 (continued)

	Age, years	Life capacity factor	Installed capacity, MW	Generated electricity, GWh	Power/swept area, W/m^2
Fife demo	5.8	0.11	7	41	
Gunfleet sands demo	6.3	0.305	12	201	1.4
Inner dowsing	10.8	0.34	97	3115	3.8
Kentish flats	14.1	0.307	90	3372	2.8
Kentish flats Extension	4.1	0.413	50	701	2.5
Kincardine demo	1.3	0.015	50	8	0
Lynn	10.8	0.34	97	3112	4.2
North hoyle	15.5	0.339	60	2733	2.1
Rhyl flats	10.1	0.336	90	2629	3.1
Scroby sands	15.1	0.306	60	2416	4.4
Teesside	4.8	0.355	62	885	5.1

Table 2.2 Generation and capacity factor data for Danish wind farms

	Age, years	Life capacity factor	Installed capacity, MW	Generated electricity, GWh	Power/swept area, W/m^2
Anholt 1	5.7	0.49	399.6	9864	2.2
Avedøre holme	8.5	0.381	10.8	306	
Frederikshavn	15.6	0.303	7.6	315	
Horns Rev I	16.2	0.412	160	9347	3.3
Horns Rev II	9.3	0.48	209.3	8213	3
Middelgrunden	18	0.252	40	1592	
Nissum bredning	1	0.395	28	97	
Nysted (Rødsand) I	15.5	0.371	165.6	8370	2.6
Nysted (Rødsand) II	8.5	0.435	207	6747	2.6
Rønland I	16	0.443	17.2	1066	
Samsø	14.5	0.391	23	1253	
Sprogø	9.2	0.335	21	566	
Tunø knob	23.6	0.302	5	313	
Vindeby (closed)	25.5	0.22	4.95	244	

Table 2.3 Declining profile of carbon intensity of electricity generation in Scotland

Year	Emissions Intensity, gCO_2e/kWh
2010	320
2011	238
2012	255
2013	216
2014	196
2015	151
2016	55
2017	24

Fig. 2.3 The declining cost of lithium-ion batteries (USD). *Source* https://www.spglobal.com/marketintelligence/en/news-insights/latest-news-headlines/as-battery-costs-plummet-lithium-ion-innovation-hits-limits-experts-say-58613238#:~:text=Mainstream%20lithium%2Dion%20batteries%2C%20now,2019%2C%20according%20to%20BloombergNEF%20data

meteorological offices charge fees. While hourly data are usually satisfactory for solar energy applications, wind data are required on a per second basis for planning and modelling large-scale wind farms. There are specialist contractors who charge large fees to measure such data using LIDAR and other techniques.

A brief selection of websites to obtain renewable energy data is provided below (accessed on October 17, 2020):

https://power.larc.nasa.gov/data-access-viewer/
https://solargis.com/maps-and-gis-data/
https://globalsolaratlas.info/
https://www.worldclim.org/data/
https://solcast.com/
http://wrdc.mgo.rssi.ru/
https://globalwindatlas.info/

It is important for the user to get an assessment of the validity of the data they use from any given website. In this respect, monthly-average wind speed data downloaded from the above NASA website is validated against measurements from Cononsyth Farm in Arbroath, Scotland (see Fig. 2.4). In late 2012 the latter farm installed

Table 2.4 The size of battery electric vehicles fleet in UK and Angus, Scotland (2019)

Car type	Model	No. of BEVs in Angus	Battery size, kWh	Storage for Angus, kWh	No. of BEVs in UK	Storage for UK, MWh	Miles/kWh
Nissan	Leaf	68	62	4216	31,016	1923	3.2
BMW	I3	18	42.2	760	8216	347	3.4
Tesla	Model 3	35	75	2625	15,876	1191	3.6
Volkswagen	Golf	13	35.8	465	9668	346	3.2
Renault	Zoe	22	52	1144	4857	253	3.7
Tesla	Model S	21	95	1995	10,085	958	3.4
Jaguar	Ipace	13	90	1170	6080	547	2.5
Nissan	ENV200	12	40	480	5860	234	2.9
Tesla	Model X	11	100	1100	5608	561	2.8
KIA	Niro	3	67	201	1191	80	3.4
Hyundai	Ioniq	3	38.3	115	1528	59	4
			Total capacity	**14,271**		**6498**	

NB With the exception of Kia and Hyundai, which use Li-polymer, all other models use Li-ion batteries

an Enercon E-30/330 kW wind turbine that records 10-min averaged wind speed at the hub height of 50 m. The manufacturer provided the following specification for the gearless machine: cut-in, cut-out, and rated wind speed of 2.5-, 34-, 12.5 m/s; diameter 33 m; maximum rotational speed of 45 RPM and 78 m/s tip speed.

The volume flow rate of air moving across a cross-sectional area A that is normal to the flow of wind may easily be shown to ρAv, where v is the wind velocity. ρ is the density of air at the given environmental temperature and pressure. The power in the wind is then,

$$P_{\text{wind}} = 0.5\rho \text{Av}^3 \tag{2.1.2a}$$

The power coefficient C_{p} is defined as the fraction of the power in the wind that is converted by the wind turbine (P_{WT}). Thus,

$$P_{\text{WT}} = 0.5C_{\text{p}}\rho \text{Av}^3 \tag{2.1.2b}$$

Using calculus it may easily be shown that the maximum value C_{p} can achieve is 16/27 which is close to 0.593. Hence even an ideal wind turbine will be, at best, 59.3% efficient.

The capacity factor of any given generator relates the actual energy conversion achieved over a given time period such as a year to the amount that would be generated if the plant operated for all hours over the latter time period. As cited above the

Table 2.5 Year 2018 electricity prices in US cents/kWh

Germany	33
Belgium	28
Italy	27
Portugal	26
Spain	24
Austria	23
United Kingdom	22
Japan	22
Sweden	21
New Zealand	20
France	19
Netherlands	18
Colombia	18
Finland	17
Poland	16
Turkey	15
United States	13
Brazil	13
South Korea	12
Canada	11
Indonesia	10
South Africa	9
India	8
China	8
Argentina	1

capacity factors for fossil fuel and nuclear power plants will be quite high as opposed to capacity factor of a wind turbine or a solar PV or thermal power plant. Data for capacity factors for wind turbines and solar PV installations are presented later on in this chapter.

For an elementary assessment of wind turbine energy generation for any given site the user may follow a three-step process:

a. Obtain data for annual- or monthly-averaged wind speed from the above-mentioned websites. Validation is provided via Figs. 2.5 and 2.6, and Tables 2.6 and 2.7 that compare on-site measured and NASA reported data.
b. Use one of the wind speed probability distribution functions (PDF) to obtain a breakdown of frequency against wind speed. A survey of such PDFs is provided by Muneer et al. [12].

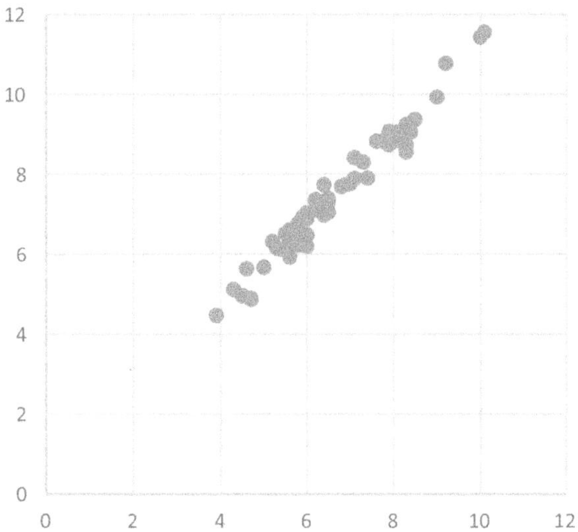

Fig. 2.4 Strong correlation between NASA estimated and on-site measured wind speed: Arbroath, Scotland (2013–17)

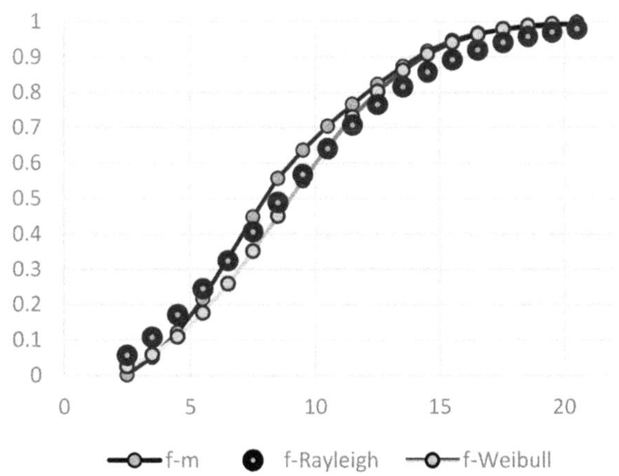

Fig. 2.5 Evaluation of models for cumulative frequency of wind speed, m/s: Cononsyth, Arbroath, Scotland. *Note* f-m: measured frequency, f-Rayleigh and f-Weibull refer to named models

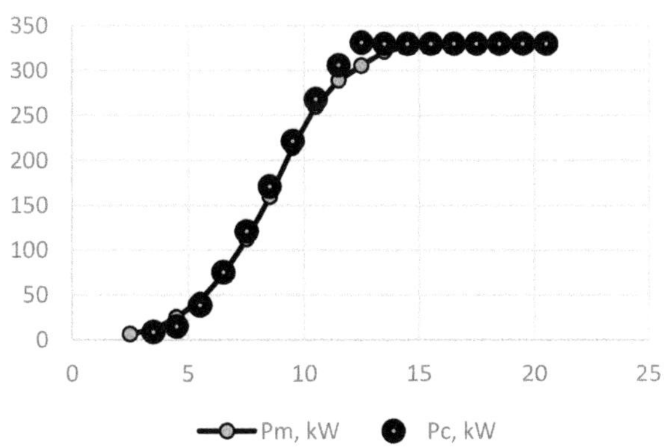

Fig. 2.6 Evaluation of wind turbine performance (*Pc*) against measured machine output (*Pm*) at given wind speed, m/s: Cononsyth, Arbroath, Scotland

Table 2.6 Monthly-averaged wind speed: Measured and NASA estimated for years 2013 and 2017

2013	J	F	M	A	M	J	J	A	S	O	N	D
Measured	8.3	6.2	6.8	7.1	6.9	4.7	4.7	5.3	6	7.1	7.9	10.1
NASA	9.3	8.8	7.8	7.2	6.8	5.7	5.7	6.6	6.6	7.8	8.2	9.9
2017												
Measured	7.3	8.2	6.5	7.4	5.4	6.5	5.8	6	5.9	8.2	7.8	8.3
NASA	9.0	8.4	8.1	6.9	6.4	5.7	5.8	6.4	7.3	8.0	8.3	9.0

Table 2.7 Measured average-annual and NASA estimated wind speed and total energy generation from Enercon E30-330 wind turbine

Year	Measured WS, m/s	NASA WS, m/s	Generation, MWh
2013	6.8	7.6	946
2014	6.5	7.3	913
2015	7.1	7.9	1035
2016	6.3	7.2	804
2017	6.9	7.7	904

c. Use the machine performance data provided by the wind turbine manufacturer to obtain monthly- and annual-averaged electricity generation. One such model was presented in Eq. (2.1.1).

Example 2.1 **Wind generation at Cononsyth Farm, Arbroath.**

Enercon E-30/330 kW wind turbine has been operating at the Cononsyth Farm, Arbroath in Scotland since the year 2012. Estimate the monthly frequency distribution of wind speed and then use that to obtain the likely electricity generation for the above wind turbine. A mathematical model of the latter machine may be easily established using information presented in Fig. 2.6 or using equation presented in Textbox 2.1 and adapting it for the above 330 kW machine. Use Rayleigh PDF for this exercise (Textbox 2.2).

Textbox 2.1 How to obtain capacity factor (CF) of a wind turbine from mean wind speed

It would be advantageous to have a simple, yet reliable method to obtain the CF from just mean wind speed, a readily available datum. This procedure may be completed via three steps:

1. Obtain mean wind speed, WS for the chosen site the local meteorological office or from the NASA website mentioned in this chapter. Tables 2.6 and 2.7 provide a comparison of measured and NASA quoted datasets as a means of validation.

2. Using the Rayleigh function (Eq. 2.1.3) obtain the probability distribution function, PDF of wind speed, WS. This procedure provides a relationship between the frequency, '*f*' and wind speed. Strictly speaking it is applicable on an annual basis. However, it also works at a monthly interval, as noted by present authors. Figure 2.5 provides a sample.
3. The mechanical energy, *dE*, converted by a wind turbine over an incremental time period $d\tau$ is given by $Pd\tau$. Thus, over a time period of, say, a month or a year the total energy, *E*, available from a wind turbine will be

$\int Pd\tau$, or $\Delta\tau \int Pdf$. $\Delta\tau$ will be the total number of hours in any given month or a year. Using normalised curves for wind turbine power, *P*, presented by Stafell* it is now possible to obtain $\mathrm{CF} = \frac{E}{\Delta\tau . P_R}$, where P_R is the rated power of the turbine (see Fig. T2.1.1), $P_R = 1$. Note that the relationship between '*f*' and WS was provided in Fig. 2.5 and between *P* and WS in Fig. T2.1.1. Equation T2.1.1 provides the mathematical relationship between *P* and WS.

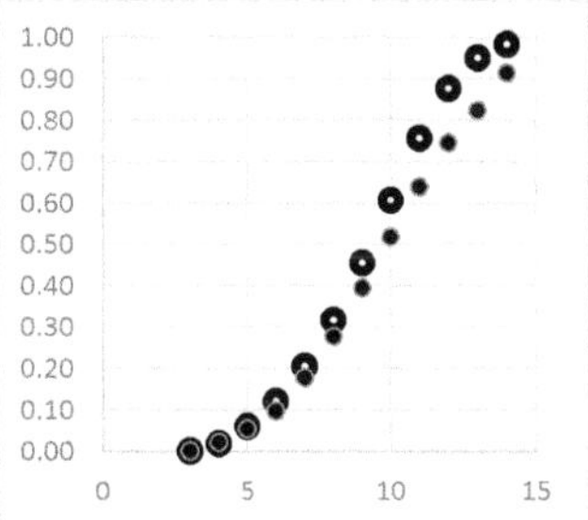

Equation T2.1.1

$P = \mathrm{Min}\ [1, -0.0014WS^3 + 0.0394WS^2 - 0.2149WS + 0.3365]$,

$3.5 < WS < 12.5$

$P = 1,\ 12.5 < WS < 25$

$P = 0,\ WS > 25$

Figure T2.1.1 Normalised wind turbine power curves

*Staffell, I, W1P5, Imperial College London, UK, March 2012.

Note: The above curves represent 16 wind turbines that range in rated power from 400 kW to 7 MW, and were produced by a number of manufacturers. All of the 16 curves fall between the lower and upper bounds represented by the smaller, solid- and bigger, hollow circles which cover the lower and upper bounds.

Using an annual mean wind speed of 7.5 m/s for Arbroath in Scotland the Capacity Factor, CF of 0.374 was obtained from above analysis. The measured CF for the period covering years 2013–2017 was reported as 0.381. The above analysis does not take into stoppages of the turbine for servicing which would obviously bring down CF.

Textbox 2.2 The Weibull distribution function for frequency of wind speed

The Probability Distribution Function, PDF of wind speed, ws or the Cumulative Distribution Function, CDF is often used for obtaining the frequency of wind speed over a given time period. There are many variations of this model that are available and are described in references provided below. Often the two-parameter model, $W(c, \delta)$ with shape parameter $c > 0$ and scale parameter $\delta > 0$ is quite adequate as shall be shown here.

$$\text{PDF}: f(ws) = \left[\frac{c}{\delta^c} ws^{c-1}\right] / \exp\left[\left(\frac{ws}{\delta}\right)^c\right]$$

$$\text{CDF}: F(ws) = 1 - 1/\exp\left[\left(\frac{ws}{\delta}\right)^c\right]$$

Pobočíková and Sedliačková, referred below, provide Method of Moments approach to obtain the scale and shape parameters:

1. Find the sample average, AVERAGE
2. Find sample standard deviation, SD
3. Estimate shape parameter, $c = 1.2785$ (AVERAGE/SD)—0.5004
4. Estimate scale parameter, $\delta =$ AVERAGE/(G(1 + 1/c)), where $G(x)$ is the gamma function. The latter function is available in popular computing packages such as MS-Excel.

Figure 2.5 provides validation of the Weibull model, evaluated against measured wind speed—frequency dataset. For this dataset the Weibull model provides a better estimate at the lower wind speeds. However, at higher wind speeds, which deliver the bulk of generate energy the two models—Rayleigh and Weibull seem to be of equal accuracy.

References

Pobočíková, I. and Sedliačková, Z., Comparison of Four Methods for Estimating the Weibull Distribution Parameters, Applied Mathematical Sciences, Vol. 8, 83, 4137–4149, 2014. [www.m-hikari.com http://dx.doi.org/10.12988/ams.2014.45389].

Lei, Y., Evaluation of three methods for estimating the Weibull distribution parameters of Chinese pine (Pinus tabulaeformis), Journal of forest science, 54, (12): 566–571, 2008.

Elmezouar, Z. C., Estimation of the shape, location and scale parameters of the Weibull distribution, https://www.researchgate.net/publoication/327721870

Teimouri, M and Gupta, A. K., On the three-parameter Weibull distribution shape parameter estimation, Journal of Data Science 11, 403–414, 2013.

Rayleigh distribution

We define a parameter in terms of monthly- or annual-averaged wind speed, Û,

$$S = \hat{U}\left(\frac{2}{\pi}\right)^{0.5} \tag{2.1.3a}$$

If Û is the monthly- or annual-averaged wind speed, then the probability of occurrence of wind speed $P(u)$ will be,

$$P(u) = \frac{u}{s^2}\exp\left(-\frac{u^2}{2s^2}\right)^{0.5} \tag{2.1.3b}$$

The cumulative distribution function is given by,

$$\text{CDF} = 1 - \exp\left(-\frac{u^2}{2s^2}\right)^{0.5} \tag{2.1.3c}$$

Note that u is the instantaneous wind speed, m/s.

Solution

Figures 2.5, 2.6 and Tables 2.6 and 2.7 present the calculation and validation of the above three steps, a–c mentioned above in the present Section. The data for measured wind speed and wind turbine generated electricity was obtained from Cononsyth Farm, Arbroath in Scotland

2.2 Solar PV

The public electric utility in Abu Dhabi which is part of the United Arab Emirates (UAE) chose a bid for construction of an ultra-large solar PV farm that will bring the cost of solar power down to 1.35 US cents per kilowatt-hour. That farm, to be completed by year 2022, will be, to date, the world's cheapest solar PV facility. The UAE is home to many of the world's biggest solar farms because of its abundant sunshine and wide, flat, empty stretches of land [13].

In October 2020, however, news emerged from Australia of the announcement of, yet again, largest-to-date solar PV plant of 10 GW capacity, to be erected in the Tanami desert, south of Darwin [14]. Example 2.2 deals with the presently proposed one such design and analysis exercise. This exercise has been developed by the present authors to demonstrate the art of solar resource modelling.

Across the globe, in the year 2019 total operational PV power capacity was 584 GW. The latter figure may be split as 580.1 GW grid connected and the rest

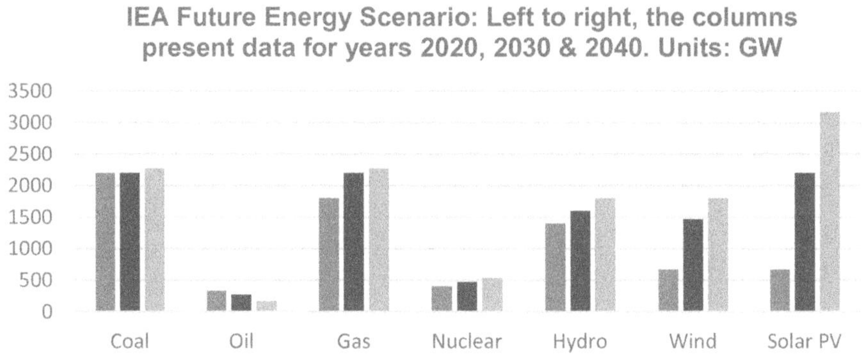

Fig. 2.7 IEA pathway to renewable energy

from off grid. Total installed renewables capacity now stands at 2564 GW, with hydropower being the dominant source at 1311 GW.

In the International Energy Agency (IEA) 'Stated Policies Scenario', solar becomes the largest source of installed capacity around the year 2035 [15]. The pathway to that scenario is shown in Fig. 2.7 which includes a rapid increase in battery storage capacity.

Table 2.8 presents data related to some of the largest PV plants across the world.

Tables 2.9 and 2.10 respectively present data from two UK farms—and solar PV capacity for worldwide plants.

Note that Table 2.9 presents data related to two farms that were commissioned in year 2017. The farms, located in a northerly country will serve as a yardstick for locations that are at lower latitudes. Their performance will only be exceeded as one approaches the equatorial belt. Of interest here are the data related to energy and power generated, the required land area, and the incident annual irradiation. Table 2.10 charts the progress of installed solar PV capacity in countries across the globe since the year 1996.

Similar to the ultra-large wind farm proposed for the North Sea, a hyper solar PV farm may be proposed for a solar PV-Hydrogen station in an arid region for sustainable energy supply to Western Europe (Fig. 2.8). A single decentralised station using a three stage compression plant will be able to supply hydrogen gas from the eastern most part of Saudi Empty Quarter to Adana, southern Turkey, a distance of 2500 km. An identical booster station would then be able to transport the gas to southern Germany, which is a further 2300 km away. Estimates show that a $9 \times 9\ km^2$ PV array area would be sufficient to provide a flow rate of 11.5 k/s of hydrogen through a 2-m diameter pipeline which translates to an annual capacity of 9 TWh [16].

Figure 2.8 and Table 2.11 provide further details.

Table 2.8 Data from some of the world's largest solar PV plants

	Project title	MW	Year	Area, km^2	Modules, Mn	Cost, Mn USD	Generation
1	Tengger Desert Solar Park, China	1547	2017				
2	Noor Abu Dhabi, UAE	1200	2019				
3	Sweihan Photovoltaic Independent Power Project, UAE	1177	2019			870	
4	Yanchi Ningxia Solar Park, China	1000	2016				
5	Datong Solar Power Top Runner Base, China	1070	2016				
6	Kurnool Ultra Mega Solar Park, Kurnool, India	1000	2017	23		1000	8 GWh on a clear day
7	Longyangxia Dam Solar Park, China	850	2018	27	4	921	
8	Enel Villanueva PV Plant, Mexico	828	2018	24	2.5	710	2 TWh/year
9	Kamuthi Solar Power Station, India	648	2016	10.1	2.5	680	
10	Solar Star Projects, USA	579	2015	12.9	1.7		
11	Topaz Solar Farm/Desert Sunlight Solar Farm, USA	550	2014	19.0	8		

NB A blank entry indicates lack of data

PV Technology

Materials, Cells, and Modules

Solar cell technologies, in general, are classified as wafer-based crystalline silicon solar cell-, thin-film- and other emerging technologies. Predominantly, there are two basic commercial PV module technologies that are in use: Wafer-based cells that are

Table 2.9 Solar PV farm data from two UK projects that were commissioned in year 2017

	Project A	Project B
Latitude	55.1	52.7
Peak DC power, MW	46	30
Peak AC power, MW	30	25.1
Land area, ha	73	53
Energy generation over 20 years, GWh	678	555
Annual-averaged irradiation, kWh/m^2-day	2.79	2.73

made from crystalline silicon either as single or polycrystalline wafers, and thin-film modules that incorporate thin layers of photovoltaic active material placed on a glass superstrate or a metal substrate using vacuum deposition. For an equal module area, thin-film materials deliver roughly half the output of a crystalline PV array.

Silicon continues to be the basic material used for the production of PV modules. Of all the semiconductor materials, the electrical, optical, and physical properties of silicon have been most rigorously researched. Other materials, such as gallium arsenide (GaAs) and cadmium telluride (CdTe), are more expensive. A disadvantage of silicon is that it is expensive to purify.

Silicon is produced in large amounts to make special steel and alloys. This metallurgical-grade silicon is obtained by reducing quartzite with coal in electric arc furnaces. Its purity is only 98–99% and embodied cost is about 50 USD per kg. The semiconductor industry purifies this metallurgical-grade silicon to a point where the relative concentration of silicon is 99.999 99%—known as seven nines concentration. Obviously high costs are associated with production of such high purity. Note that electronic grade silicon is even purer with the purity target being nine nines, i.e. 99.999 999 9%.

Solar silicon cells are typically made from single, or monocrystalline, polycrystalline, or amorphous solids. Monocrystalline silicon is the most efficient because the crystal is free of grain boundaries, which are defects in the crystal structure caused by variations in the lattice that tend to decrease the electrical and thermal conductivity of the material. Polycrystalline silicon has grain boundaries, i.e. the portions of single crystals are even visible to the naked eye. Amorphous silicon (a-Si) is the non-crystalline form of silicon where the atoms are arranged in a relatively haphazard way. Due to the disordered nature of the material, some atoms have a dangling bond that disrupts the flow of electrons. Amorphous silicon has lowest power conversion efficiencies of the three types but is the least expensive to produce.

Single Crystal or Monocrystalline Silicon Cell

The work horse of PV industry has always been the crystalline silicon cell, fabricated from a single crystal or cast polycrystalline silicon that is sliced into wafer of about 10 × 10 cm area and 350 μm thickness. These cells typically show efficiencies between 13 and 15%, although in recent years there has been a concerted effort to increase

Table 2.10 Installed solar PV capacity (GW) in countries across the globe in given years

Year 1996	Year 2019	Country/Region
0.01	7.24	Africa
0	0.45	Algeria
0.08	346.44	Asia Pacific
0.02	15.93	Australia
0.01	1.58	Austria
0	4.53	Belgium
0	2.48	Brazil
0	1.07	Bulgaria
0.01	3.31	Canada
0	2.65	Chile
0.01	205.49	China
0	1.7	CIS
0	2.07	Czech Republic
0	1.08	Denmark
0.01	1.67	Egypt
0.07	146.67	Europe
0.01	2.06	Europe (other)
0.01	10.57	France
0.03	48.96	Germany
0	2.76	Greece
0	0.51	Honduras
0	1.28	Hungary
0	35.06	India
0	1.44	Israel
0.02	20.91	Italy
0.06	61.84	Japan
0	1	Jordan
0	0.88	Malaysia
0.01	4.44	Mexico
0	5.58	Middle East
0	0.74	Morocco
0.01	6.73	Netherlands
0.03	70.05	North America
0.01	1.32	Other Africa
0	7.34	Other Asia & Pacific
0	0.63	Other CIS

(continued)

Table 2.10 (continued)

Year 1996	Year 2019	Country/Region
0	1.26	Other Middle East
0	3.11	Other S. & Cent. America
0	1.33	Pakistan
0	0.92	Philippines
0	1.3	Portugal
0.01	0.83	Romania
0	1.06	Russia
0	1.39	Slovakia
0	8.75	South & Central America
0	3.06	South Africa
0	10.51	South Korea
0.01	11.06	Spain
0.01	0.64	Sweden
0.01	2.52	Switzerland
0	4.15	Taiwan
0	2.99	Thailand
0	6	Turkey
0	1.88	UAE
0.01	13.4	UK
0	5.94	Ukraine
0.01	62.3	USA
0.17	**586.42**	**World**

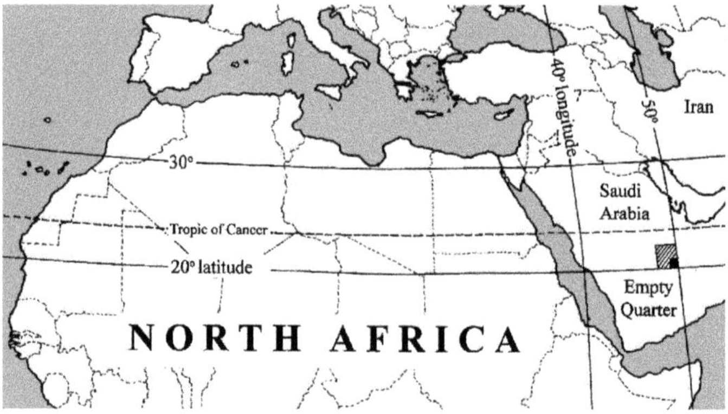

Fig. 2.8 Ultra-large solar PV plant for Saudi, empty quarter

Table 2.11 Arid/semi-arid locations around the globe with potential for installation of Hyper PV stations

Energy markets	Arid/semi-arid location or desert
USA, Canada	Nevada, Baja
S America	Atacama (Chile)
OECD-Europe	Southern Spain
Middle East, OECD-Europe	Saudi Empty Quarter
China, CIS states	Gobi
India	Rajasthan
N Africa	Sahara
S Africa	Namibia, Kalahari
Australia	Great Victoria, Gibson, Tanami

that figure. Reliable manufacturers of this type of PV module offer guarantees of up to 20–25 years at 80% of nameplate rating.

Polycrystalline or Multicrystalline Silicon

These cells are made up of various silicon crystals formed from an ingot. They are also sliced and then doped and etched. They demonstrate conversion efficiencies lower than those of monocrystalline cells. Good manufacturers typically guarantee polycrystalline PV modules for 20 years. Crystalline silicon PV modules are expected to remain a dominant PV technology for the near future, with a forecasted market share of about 50% by that time. This is due to their proven and reliable technology, long lifetimes, and abundant primary resources.

Thin Film (Non-silicon)

The high cost of crystalline silicon solar cells is due to the cost of Si wafers, which constitute 40–50% of the cost of the finished modules. Therefore, thin-film technology was launched with the objective of reducing the cost of the material by using lesser quantity of absorbing material. The selected materials are all strong light absorbers and only need to be about 1 μμ in thickness. Thus the costs are significantly reduced. Some of the most attractive materials used for thin-film solar cells are amorphous silicon, cadmium telluride, copper indium diselenide, and copper indium gallium diselenide. Thin-film modules are constructed by depositing extremely thin layers of photosensitive materials onto a low-cost backing such as glass, stainless steel, or plastic. Thin-film solar cells consist of layers of active materials about 10 nm thick. Commercially available thin-film PV has not attained field efficiencies greater than 10%, as compared with the 16–18% efficiency of crystalline silicon PV modules.

Amorphous Thin Film

The first thin-film solar cell produced was amorphous silicon (a-Si). Amorphous thin films have evolved from an efficiency of 2–5% to above 12%. Stability concerns have, however, been reported for this technology. Changes in performance after exposure to light are well known and the reported efficiencies seem to be those measured before any light-induced changes occur. Based on early a-Si single junction cells, amorphous tandem and triple cell configuration have been developed. To reach higher efficiencies, thin amorphous and microcrystalline silicon cells have been combined to form micromorph cells. Another option is the combination of single-crystalline and amorphous PV cell technology. The HIT or HJT (heterojunction with intrinsic thin layer cells) technology is based on a crystalline silicon cell coated with a supplementary amorphous PV cell to increase efficiency.

The basic device feature of the HJT cells is that on the front (illumination) side successively an intrinsic amorphous silicon passivation layer and a p-doped amorphous silicon emitter are both deposited by plasma enhanced chemical vapour deposition. On top of the silicon layers, an antireflective transparent conductive oxide (TCO) is deposited by physical vapour deposition, and the charge collection is made by a screen-printed metallic contacting grid. On the back side, an electron collecting stack is used, and it is also composed of amorphous silicon layer, a doped *n*-type amorphous silicon, a TCO layer, and a metallic contacting layer.

HJT combines the best qualities of crystalline silicon with those from amorphous silicon thin film to produce a high-power hybrid cell that delivers high performance. The 2019 "International Technology Roadmap for Photovoltaic" report expects HJT cells to gain a market share of 12% in 2026 and 15% by 2029. The HJT production concept was developed by SANYO Electric in the 1980s [17].

The temperature coefficient is the main performance indicator of a solar module under real, in-situ conditions. All module manufacturers quote their power output at 25 °C module temperature. The temperature coefficient informs the modeller/designer of the drop in output when the module temperature is raised under the influence of irradiance. The temperature coefficient of Sanyo HJT modules is −0.258%/°C.

Table 2.12 presents the data sheet for one manufacturer of heterojunction technology (HJT) solar PV modules.

Efficiency

The performance of a solar cell is measured in terms of its conversion efficiency at converting sunlight into electricity, i.e. the efficiency of a PV device is defined in terms of the power produced from the incident photons. The common approach is to obtain the light-generated current–voltage (I–V) characteristics and ascertain the maximum power point. The standard measurement and reporting conditions are defined by reference to spectral irradiance, total irradiance, temperature, and area of the device. The following standard test conditions are used when quoting cell efficiencies for terrestrial applications: global irradiance = 1000 W/m^2, AM (air mass) 1.5, and a temperature of 25 °C.

PV modules, strings, arrays, and inverters

A PV module or panel is a grouping of PV cells. The voltage generated by a single PV cell is inconveniently low. Several cells are always joined in series so that their voltages add up to a more useful value. The series connection of cells forms a unit

Table 2.12 Data sheet for Jinergy heterojunction technology (HJT) JNHM72-410 solar PV module

Mechanical parameters	
Cell (mm)	156 × 156 HJT
Module dimensions (LxWxH), mm	1968 × 992 × 6/5
Weight, kg	26.7/22.6
Glass thickness, mm	2.5/2.0
Number of cells	72 (6 × 12)
Number of diodes	3
Electrical parameters	
Maximum voltage, DC V	1500
Temperature cycling range, Celsius	−40 to 85
Maximum series fuse rating, A	15
Maximum wind/snow load, Pa	2400/5400
*Data for Bifacial modules under STC**	
Maximum power, Pmpp (W)	420
Maximum power voltage at BSTC, Vmp (V)	44.93
Maximum power current at BSTC, Imp (A)	9.35
Open circuit voltage, Voc (V)	50.48
Short circuit current, Isc (A)	9.83
Temperature coefficient of power	-0.35%/K
Efficiency at STC	20.4%
*Data for Monofacial modules under STC**	
Maximum power, Pmpp (W)	380
Maximum power voltage at BSTC, Vmp (V)	44.82
Maximum power current at BSTC, Imp (A)	8.5
Open circuit voltage, Voc (V)	50.01
Short circuit current, Isc (A)	8.84

*Standard Test Conditions (STC): 1000 W/m^2, 25 °C, AM = 1.5

called a PV module. The front consists of a window of low-iron glass with high transmission characteristics, which protects the surface of the PV material.

A string is a group of modules that are wired in series. This is to increase the voltage as modern solar electric DC systems operate at a minimum of 48 V. High-voltage grid-tied systems produce up to 600 V.

One of the attractions of PV modules is that connecting modules to form array may increase the power rating of a system.

The inverter is an essential component of a PV system as it is responsible for the effective conversion of the variable DC output of the PV modules into clean, sinusoidal AC current with the required frequency of either 50 Hz or 60 Hz. The optimal sizing of an inverter is dependent on the output from the PV generator which is controlled by the local climate, the surface orientation, and inclination of the modules. PV modules have temperature coefficients of power that affect the power output which is dependent on solar radiation, cell temperature, and the solar spectrum. The cell temperature, however, rises as the intensity of irradiance increases thereby limiting the optimum performance of the modules below the nominal DC power. A reduction in the intensity of the radiation that reaches the surface of the module reduces its output power below its rated capacity thereby causing the inverter to operate under part-load conditions with reduced system efficiency if the inverter is not properly sized. For optimum performance of a PV system, the rated capacity of the inverter should be higher than the rated capacity of the system in order to prevent operations at overload conditions. Improper sizing of inverters thus increases the total energy costs. The selection of an inverter is dependent on its optimal features such as efficiency, frequency regulation, low standby losses, power-correction factor, servicing, cost, and reliability. The inverter's modularity facilitates continuous scaling of its capacity to optimise energy for all conditions including lower irradiance conditions. One distinct feature of the high yield inverter is its ability to automatically reconfigure itself for continuous power generation in the event of any module inverter trip. It also has inbuilt mechanisms that rotate the standby and active duty of each module in order to share generating capacity thereby extending the service life of the inverter. At lower solar irradiation level, the inverter automatically resizes its modular solution to match the power output of the modules [18].

Cabling

Power losses in direct current (DC) systems are due to voltage drops. Care must be taken when selecting the size of the cable to be used. If a smaller cable is used then the voltage drop increases and in an off grid or stand-alone system this can have a major impact on the system where the battery voltage is lower than expected. Voltage drop can be calculated as follows: $V = I \text{ x } R$, where V is the voltage drop in the cable, I the current, and R the resistance of the cable in ohms. Note that the electrical resistance depends on the length and cross-sectional area of the cable. Table 2.13 presents data related to the electrical resistance of the cable.

Table 2.13 Electrical resistance of cable

Conductor cross sectional area (mm^2)	Resistance (Ω/m)
2.5	0.0074
4	0.0046
6	0.0031
10	0.0018
16	0.0012
25	0.00073
35	0.00049

As an example, the voltage drop in a 100 m long cable that has a 10 mm^2 cross-sectional area, carrying a current of 20 A will be 20 × 0:0018 × 100 = 3.6 V.

Recent developments to increase the output from PV modules

Earlier in this section, we have discussed successful attempts to increase the efficiency of PV cells via HJT technology that combines crystalline silicon with amorphous silicon thin film to produce high-power hybrid modules that deliver high performance. Other ways in which technology has moved on is to develop bifacial modules and augmentation of irradiation by using surfaces of high reflectance in the near vicinity of PV modules.

Bifacial modules

The vast majority of the currently installed PV modules are mono-facial. These modules only convert irradiation on the front of the module into an electric current. However, bifacial modules—those that have PV cells on front and back of the module—are increasingly being developed and used in solar farms. Some manufacturers claim that bifacial modules can generate up to 30% more energy. However, that will depend on a number of factors such as type of cells used, the angle of tilt of the module, the reflectance and type of horizon, the view factor, and the solar climate. A low tilt angle of the PV module will ensure a higher sky view factor while decreasing the ground view factor. However, that would also increase the view factor of the back panel of the module. Thus, in any given farm layout an optimum design may be arrived at by using the module tilt angle as a variable.

In March 2019 a joint Energy Technology Partnership (ETP), industry and academia led doctoral research programme was initiated at Heriot-Watt University. The participants were ETP, Wood Group—an international solar and wind energy consultancy, Heriot-Watt, and Edinburgh Napier University. Dr. M Gul of Heriot-Watt and the present author, Professor T Muneer jointly supervised the doctoral candidate M Alam who has extensively monitored the following regime of experiments related to bifacial gain of solar energy. Figure 2.9 shows the experimental setup.

Three surfaces, soil, concrete and white tiles, with albedo ranging 0.2–0.6 were placed around the front- and back-facing modules. For each of the above surface albedo, the lower edge of the rear of the PV modules was kept at a height of 0.5- and 1 m. Results show that for all surfaces there is a strong element of solar altitude dependence on the bifacial gain factor, particularly for the case when the modules are placed close to ground. The bifacial gain factor is presently defined as the ratio of total incident energy on the front and back of the module to the frontal irradiance. For example, that latter ratio for soil and concrete varies from 1.1 to 1.25 and 1.1 to 1.5 respectively, as the sun passes from sunrise to noon. Those are mainly diffuse surfaces. For specular surfaces such as white tiles, the gain is considerably higher and may achieve ratios that are double than that of the latterly mentioned surfaces. The case when the height of the lower edge was 0.5 m showed a stronger dependence of bifacial gain on solar altitude with the gain also being higher for like-with-like scenario. That is primarily because the view factor is higher for lower height of the

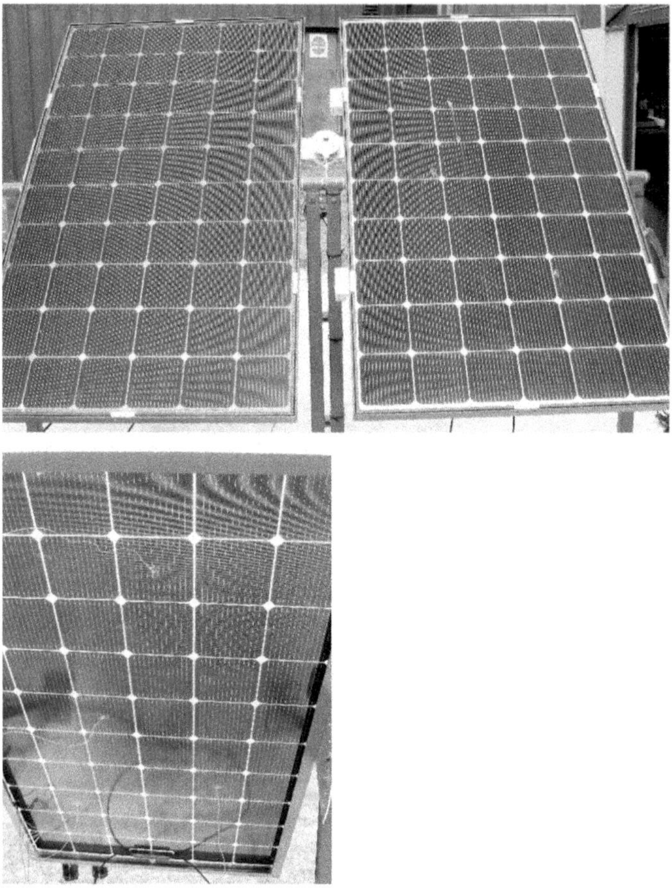

Fig. 2.9 Experimental investigation of bifacial solar PV gain: top picture (front side of the PV module with Kipp and Zonen pyranometer), and bottom (rear side of the bifacial module)

module, e.g. the back side of the module, shown at the bottom end of Fig. 2.9, will have a view factor (VF) of 0.854 while the VF of the front of the module is only 0.146 for infinite horizon. Note that the tilt angle of the module was 45°.

Radiation augmentation

There are three main components of solar radiation on an inclined surface: (a) Direct beam; (b) diffuse from the sky; and (c) reflected from the foreground, where the reflected component is generally 10% of the total incident radiation.

In general, ground albedo, or simply 'albedo', is defined as the ratio between the ground-reflected radiation and the global radiation incident on the ground. It is a dimensionless fraction and is measured on a scale from 0 to 1. An albedo of 0 is of a perfectly black surface (no radiation is reflected, all absorbed), and an albedo of 1 is of a highly reflective material (a highly polished mirror) possessing total reflection

Table 2.14 Reflectance recorded for each material under varying sky conditions

	Overcast		Part-overcast		Clear-sky	
	Reflectance	SD	Reflectance	SD	Reflectance	SD
Grass	0.19	0.03	0.19	0.02	0.21	0.02
Sand	0.35	0.05	0.37	0.05	0.36	0.07
White pebbles	0.55	0.05	0.56	0.05	0.58	0.04
Cement slabs	0.13	0.04	0.16	0.03	0.17	0.04
White tiles	0.7	0.02	0.7	0.02	0.7	0.06
White board	0.59	0.04	0.6	0.08	0.65	0.03
Aluminium foil	0.73	0.05	0.75	0.05	0.72	0.08

(no radiation absorption). Ground reflected radiation may reach up to 100 W/m^2. In the absence of ground reflected radiation measured data, a ground albedo value of 0.2 is widely accepted and is used in the modelling of PV systems. Foreground surfaces may have different albedos, and hence a constant albedo value is unsuitable. The albedo value becomes even more significant in the case of bifacial PV cells, which are meant to produce more energy as they collect radiation on the front as well as on the rear side. Overcast skies are predominant in northerly locations and will have a high frequency of low solar altitudes, and thus ground reflection is an important contributor towards total solar energy gain.

Table 2.14 presents experimentally determined data on the reflectance of surfaces that may be used for enhancing PV module irradiation for boosting the energy output. Further details are available in the work of Gul et al. [19].

The front and back of any given PV module will 'see' a horizon that may belong to one of the following categories:

a. Infinite, uniform horizon
b. Infinite, non-uniform horizon
c. Finite, uniform horizon
d. Finite, non-uniform horizon.

With the exception of above case 'a' the estimation of ground-reflected radiation is ideally handled by means of a finite-element approach with each element having its own, given reflectance. Muneer et al. [20].

Resource assessment and elementary modelling: solar energy

As compared to wind, solar energy modelling is much more involved. To start with the user is dealing with spherical trigonometry regarding sun's position calculation and its movement in the sky hemisphere. The measurement of irradiance also requires specialist equipment such as pyranometers (to measure global and sky-diffuse irradiation) and pyrheliometer (to record solar-beam irradiance). The sky-diffuse pyranometer requires almost daily adjustment of the shade ring, then the correction factor itself has been the subject of a number of researches.

One solution is to use proprietary software and there are a number of them that are available in the marketplace. One popular software is HOMER. This software has the ability to simulate and optimise stand-alone and grid-connected power systems comprising any combination of wind turbines, PV arrays, run-of-river hydropower, biomass power, internal combustion engine generators, fuel cells, batteries, and hydrogen storage, serving both electric and thermal loads. It has the ability to handle electricity and heat sectors, leading on to design autonomous and part-load renewable energy systems.

It can generate large number of outputs which will lead to selecting technology options, perform technical feasibility, assess available energy resources and economics.

To perform successful simulations the following information is required: Resources, loads, equipment costs, and their performance. HOMER can generate reports to contain:

- Simulation and analysis of varying loads and resources such as wind and solar energy
- Optimisation
- Sensitivity Analysis.

HOMER can be accessed at https://www.homerenergy.com/.

If the user wishes to avoid the 'blackbox' approach and have complete control of all modelling streams then they will have to resort to the fundamentals of solar radiation modelling. Muneer [21] has provided a large suite of FORTRAN and EXE files that enable the user to handle many aspects of solar radiation. The total number of FORTRAN files presented in the latter reference is 43. Some of the key routines presented enable calculation of:

- Solar time from clock time and vice-versa
- Solar declination angle
- Sun's geometry
- Hourly, daily, monthly, and annual clear- and all-sky irradiation for any location
- Frequency of occurrence of any given threshold of radiation and daylight illuminance
- Slope radiation
- Quality control of measured solar and illuminance data.

Example 2.2 shall demonstrate the type of computations that can be performed using the above-mentioned routines.

Example 2.2 **Design of a Large Solar PV Plant**

Sun Cable's $22 billion projects is related to building the world's largest, 10 GW solar and battery power station, as well as the world's largest sub-sea transmission link at a 12,000 ha site that is about 750 km south of Darwin. The aim is to connect the energy plant to Singapore via an undersea cable [14].

The company says that the chosen site was attractive for the balance it struck between solar resource and transmission distance. The chosen site will probably be closely located to railway track and also have access to highway. The 10 GW solar plant and the associated battery storage facilities of up to 30 GWh will use a high-voltage direct current sub-sea cable of 3750 kms to provide solar power across to potential customers in Singapore.

Note that if transport, production, and processing of materials are all taken into account brown coal, hard coal and natural gas respectively emit 1.183-, 1.142- and 0.572 $kgCO_2/kWh$ of electricity.

The year 2018 annual electricity generation in Singapore was 50.4 TWh and it has 1.4 million households. The respective electricity tariff for domestic and business customers, excluding taxes are 0.2143 and 0.155 USD/kWh. Note that the global average price of electricity is 0.14 USD/kWh. Currently, 95% of the electricity used in Singapore is generated from burning natural gas with a carbon intensity of 0.572 $kgCO_2/kWh$ as mentioned above.

In view of above data, it may be assumed that the bulk price of electricity sold to Singapore market may be half of the charge that is passed on to the domestic and business customers, i.e. an average cost of 0.09 USD/kWh.

Calculate:

a. Number, tilt, and spacing of PV modules of a suitable design.
b. Project cost
c. Monthly-averaged energy generation of the plant
d. Year 2025 cost of the plant and the electricity transmission cost to Singapore
e. Carbon dioxide savings when compared to coal-fired or natural gas power plants.

Solution

Let us assume that the Jinergy JNHM72-410 monocrystalline PV modules are selected for this project. The data for these modules is provided in Table 2.12. The November 2020 cost is quoted as USD 160 for a single module.

The peak design power of the plant is 10 GW, so we need 23.8 million modules which would cost a total of USD 2.286 billion assuming a 40% discount for bulk purchase. As module costs account for 50% of the total cost, the project cost would thus be USD 4.571 billion. The current operation and maintenance (O&M) cost for PV plants is USD 4–9.5/kWDC/year. The total O&M cost for this plant would thus be 70 million USD per year. Thus, the total annual cost of the project that includes capital and running expenditure would be close to 300 million USD, assuming a 20 year project life.

The ideal tilt (slope) of the module is often determined by the $0.9 \times$ latitude rule. The location of the project is given as 750 km south of Darwin which would be in the Tanami desert of Australia. The presently chosen latitude and longitude are 23° South and 131° East. The ideal tilt would thus be 21°. However, in the interest of reducing shading from adjacent rows of PV arrays, it would be advisable to adapt a much lower tilt of say, 15°. An inter-row PV string spacing of 3 m could be adopted. This means the ground area of 5×1 m will be used up for each module. Thus in

Table Ex2.1 Data obtained from NASA database

Month	Global radiation[a]	T_{max}, °C	T_{min}, °C	$E^b_{generated}$
1	7.29	26.3	25.0	2.64
2	6.54	26.5	25.2	2.50
3	6.15	26.7	25.3	2.49
4	5.00	26.1	24.8	2.16
5	4.11	25.0	23.6	1.94
6	3.64	24.1	22.7	1.81
7	3.93	23.4	22.0	1.92
8	4.73	23.0	21.7	2.15
9	5.61	23.0	21.6	2.33
10	6.27	23.3	22.0	2.48
11	6.64	24.3	23.0	2.48
12	6.62	25.4	24.1	2.40

[a] kWh/m^2-day
[b] Wh/module-day

12,000 ha site 24 million modules may be accommodated which is very close to the 23.8 million figure obtained from power delivery standpoint.

Using the link for NASA database provided in Sect. 2.1. The monthly-averaged global irradiation, maximum and minimum temperature are downloaded and those data are presented in Table Ex2.1.

From this point onwards the calculations become more involved. Solar geometry and irradiation computations require specialist knowledge as we are dealing with spherical trigonometry. The reader is referred to books on solar radiation. One such text which additionally provides software for these computations is due to Muneer [21]. The chain of calculations is provided in Table Ex2.2. Only two months data is shown by way of example. The main steps are: Disaggregate the daily-horizontal into hourly irradiation and likewise obtain hourly ambient temperature from daily maximum and minimum values obtained from NASA website, split hourly irradiation into the beam and diffuse components, obtain slope radiation which leads to PV cell temperature which will provide the varying efficiency of the PV module. The hourly power delivery from the installation is thus obtained.

Note that the total annual energy delivered is 830 kWh/module which translates into 19.8 TWh for the plant. If 90% of this energy was exported to Singapore that would satisfy 35% of the demand.

We now undertake the analysis of cost associated with the transmission of electricity from Australian solar power plant to the Singapore market. Refer to Sect. 5.3, part (ix). A cost of US cents 0.07/kWh/thousand km was quoted. That would translate into 12.5 million USD per year. Thus, the overall cost that includes capital, O&M, and transmission would be 312.5 million USD per year. On the other hand, the annual revenue from the sale of electricity would be 1604 million USD. Indeed a profitable enterprise.

Table Ex2.2 Hourly calculation of horizontal global, diffuse and slope radiation (W/m^2), ambient and cell temperature (Celsius), and module power output

Hour	IG	ID	ISlope	Ta	TCell	P, W
January						
5.5	16	0	0	25	25	0
6.5	170	89	125	25	28	52
7.5	354	177	310	25	33	126
8.5	549	262	507	25	39	203
9.5	727	336	690	26	44	271
10.5	865	392	832	26	47	322
11.5	940	421	910	26	50	349
12.5	940	421	910	26	50	349
13.5	865	392	832	26	48	322
14.5	727	336	690	26	44	270
15.5	549	262	507	26	39	202
16.5	354	177	310	26	34	126
17.5	170	89	125	26	29	52
18.5	16	0	0	26	26	0
June						
5.5	0	0	0	23	23	0
6.5	0	0	0	23	23	0
7.5	101	55	155	23	27	64
8.5	246	125	319	23	31	131
9.5	388	189	481	23	36	194
10.5	503	237	611	24	40	244
11.5	566	263	683	24	42	270
12.5	566	263	683	24	42	270
13.5	503	237	611	24	40	243
14.5	388	189	481	24	37	194
15.5	246	125	319	24	32	131
16.5	101	55	155	24	28	64
17.5	0	0	0	24	24	0
18.5	0	0	0	24	24	0

On the environment front, the annual carbon dioxide emissions from generation of 19.8 TWh of electricity may be estimated thus; from hard coal- 22.6 and natural gas 11.3 kilo tonnes. The PV electricity will also have Carbon footprint, but in light of material presented in Sect. 6.1 per kWh, carbon dioxide emission would be around 20 g/kWh$_e$ adjusted for the Australian solar climate. The solar PV related carbon dioxide emission would thus be less than 0.4 kilo tonnes.

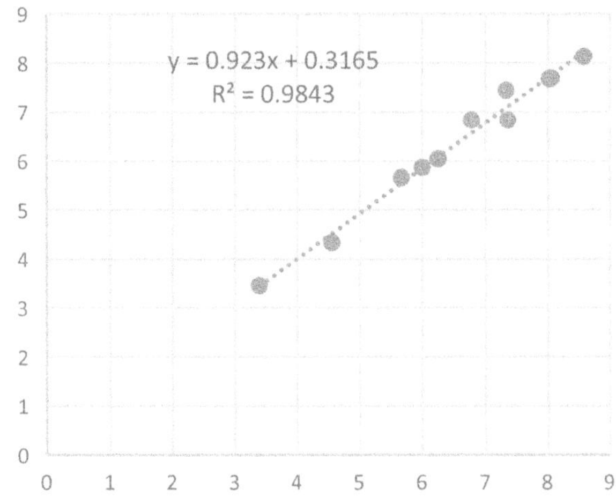

Fig. Ex2.1 Strong correlation between NASA estimated and on-site measured solar radiation: Bahrain (2001)

It is worth mentioning that the capacity factor for the proposed Australian plant works out at 22.6% which is extremely healthy when one compares to the figures for three UK projects: 8.5 and 10.6% (Table 2.9), 10.3% for solar meadows project near Edinburgh (Sect. 6.3), and 10.4% for Arbroath solar farm (Sect. 6.5).

The areal density of energy delivery is also quite healthy for the Australian plant with a figure of 165 kWh/m^2-year. In contrast, the figure for solar meadows project is 28 kWh/m^2-year. The big difference is due to the combined effect of two factors that work in favour of the former: high solar income and low latitude that enables closer spacing of the PV arrays. For higher latitudes shading between arrays has to be avoided by selecting low tilt angles and larger spacing.

For the sake of validation of data sources from the NASA website, Fig. Ex2.1 and Table Ex2.3 have been created wherein NASA data has been compared against five-minute averaged measurements that were carried out in Bahrain for a period of about two years in a research project. That project was funded by the University of Bahrain, the Bahrain Meteorological Office, and Edinburgh Napier University.

2.3 Solar Water Heating

Passive solar energy research and development precedes the innovation that has taken place with regard to solar PV. This section presents material on those solar water heating technologies that are proven, mature and economical. Of all the applications of solar energy, thermal technologies are the oldest and most mature. They are also most economical. From very early architecture of human civilisation, one notes the use of passive solar design concepts with the use of southern facades for maximising solar gains for northerly locations, while use of thick walls kept heat out in locations that were known for their hot summers. Likewise, at the turn of the twentieth century,

Table Ex2.3 Measured and NASA-estimated solar radiation at Bahrain

Month	Global, measured	Diffuse	Direct	North	East	South	West	Diffuse fraction	Direct fraction	NASA global, 2001	NASA global, 1984–2019
1	4.25	0.58	1.75	4.78	2.64	1.39	2.86	32.7	67.3	3.86	3.61
2	4.89	0.75	2.14	4.19	2.58	1.78	2.31	32.4	63.6	4.59	4.40
3	6	0.86	2.42	3.47	3.11	2.08	3.92	34.7	65.3	5.86	5.17
4	7.36	1.17	3.06	2.58	3.53	2.25	5.11	30.6	69.4	6.83	5.99
5	8.06	1.69	3.25	1.75	3.61	2.44	5.61	30.3	69.7	7.69	7.00
6	8.58	2.06	3.44	1.31	3.94	1.81	6.78	21	79	8.13	7.62
7	8.03	1.86	3.25	1.58	3.64	2.31	5.72	28.7	71.3	7.68	7.23
8	7.33	1.39	2.97	2.22	3.42	2.44	4.89	33.3	66.7	7.44	6.94
9	6.78	0.97	2.86	3.25	3.42	1.94	4.83	28.7	71.3	6.83	6.44
10	5.67	0.72	2.53	4.53	3.19	1.47	4.19	26	74	5.66	5.33
11	4.56	0.64	2.06	4.97	2.72	1.31	3.25	28.7	71.3	4.33	3.96
12	3.39	0.61	1.56	3.97	1.97	1.53	1.86	45	55	3.45	3.39
Annual	**6.25**	**1.11**	**2.61**	**3.22**	**3.14**	**1.9**	**4.28**	**31**	**69**	**6.04**	**5.59**

there were over a hundred thousand solar hot water 'pillows' that were in use in Japan. These 'pillows' were placed on rooftops to collect solar energy during the day and the hot water was then used for evening bathing. Similarly, there was a booming solar water heating industry in Florida that peaked around the year 1905, the decline of that industry being a direct consequence of cheap oil. However, with the rapidly rising fossil fuel prices and increasing awareness of the carbon impact, there is now a concerted effort at global, national, and regional levels for energy conservation.

2.3.1 Markets and Uptake

Since 1980 the use of solar technology has increased at the rate of about 30% yearly [22]. Among all devices utilising solar energy, domestic hot water systems are the most frequently used.

In 2010, Renewable Energy Policy Network has reported that, worldwide, 70 million houses were using solar water heating systems [23].

Solar water heating systems are becoming wide spread and are now contributing significantly to both domestic and industrial sectors in several countries. China currently dominates the global solar thermal market. Chinese companies manufactured 28 million square metres of system in 2009 which was above 80% of global solar hot water/heating output [24].

Apart from China, Germany, Turkey, Brazil, and India lead the solar hot water market. Germany marked a record in 2008 with an estimated solar water heating capacity of about 1.1 GW.

Similarly, the government of Brazil targeted a reduction of more than 1000 MW additional energy installations by the year 2015 through the initiation of solar water heating systems.

On average, about 20,000 solar water heating systems are being installed each year in India. On the other hand, the US domestic market capacity is still relatively small compared to the rest of the world but it is increasing. Solar water heating market is also expanding in the countries of Africa, such as Ethiopia, Kenya, South Africa, Tunisia, and Zimbabwe. Palestine has the highest installation capacity among the countries of Middle Eastern and North African regions. The capacity accounted for 68% of all the households that installed SWHs routinely in new buildings in that region [24].

Although solar water heating systems have a good prospectus, only a fraction of the projected possible utilisation has been realised. Water heating typically requires a temperature build-up of up to 60 °C.

By the end of year 2017, the world had an installed solar thermal capacity of 474 GWth delivered from a collector area of 676 million m^2, operational across 68 countries. The collectors used were unglazed water-, flat-plate-, evacuated-tube- and unglazed and glazed air collectors.

The leading markets were: China (334.5 GWth) and Europe (54.3 GWth), which accounted for 82% of the total capacity. The United States and Canada (18.9 GWth),

Asia, other countries (13.3 GWth), Latin America (13.6 GWth), the Middle East and North Africa, MENA (Israel, Jordan, Lebanon, Morocco, Palestine and Tunisia, 7.0 GWth), Australia and New Zealand (6.7 GWth), and Sub-Saharan Africa (1.6 GWth).

Evacuated-tube collectors share was 71%, followed by flat-plate collectors (23%), unglazed water collectors (6%), and glazed and unglazed air collectors (0.3%).

The countries with the lion's share of solar thermal installations on a per capita basis were: Barbados, 540 W; Cyprus, 440 W; Austria, 413 W; Israel, 397 W; Greece, 300 W; Australia, 276 W; Palestine, 269 W; China, 242 W; Turkey, 201 W and Denmark, 199 W.

Note that with the gas-fired boilers there is a significant environmental penalty associated with the combustion of gas with 14 kg of CO_2 and 65 g of NOx emitted per GJ of energy released.

2.3.2 *Hot Water Draw-Off Profiles*

In this section, information is provided on the hot water draw-off profiles, the classification of solar water heating systems, solar collector design and performance, and economics.

Figures 2.10 and 2.11, respectively show typical draw-off profile of hot water for a representative dwelling in the UK and the associated energy needs. See Reference [25] for further details.

Disparate needs and habits of individuals give rise to variations in hot water demand for households in any country. Currently, there is a dearth of published information on water consumption patterns. Water companies do undertake periodic research on this subject but their accounts are not easily available.

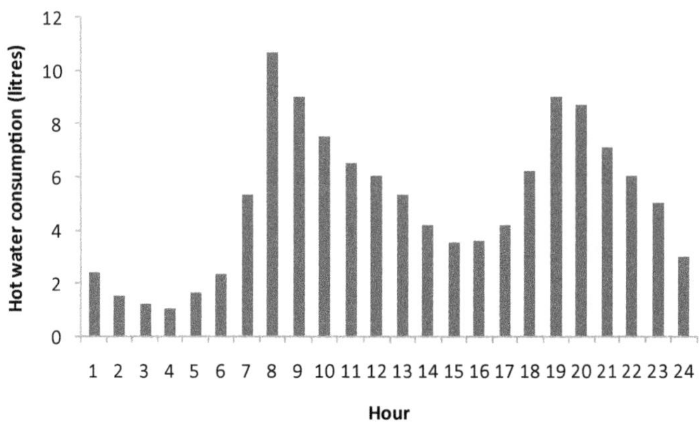

Fig. 2.10 Hot water use—UK dwellings

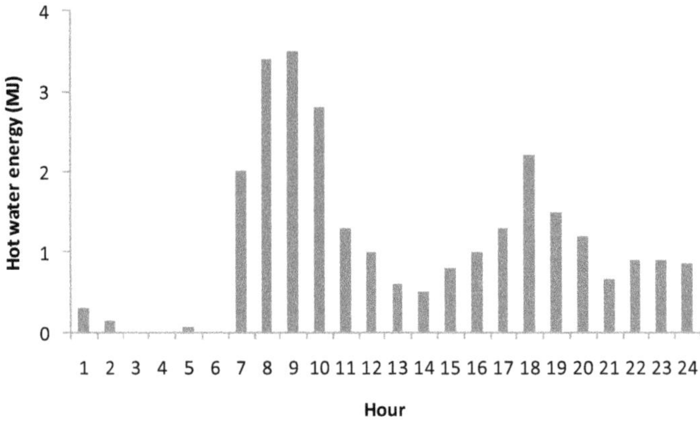

Fig. 2.11 Energy delivered to hot water—UK dwellings

In the UK every person uses approximately 150 L of water a day, a figure that has been growing every year by 1% since 1930. The average water consumption by Spain households was 142 l/person/day [26].

The required temperature of hot water delivered at the tap for bathing and hand washing is 40 °C. However, dishwashing and laundry respectively require temperatures of 50 °C and 80 °C. It is therefore usual to store hot water at 60 °C with a temperature top-up depending on the application of use. The rule of thumb for solar water heating system design is to use 80 l/m^2 collector area for tropical locations, 60 l/m^2 for the US, Central Europe, and Canada, 40–45 l/m^2 for England and Wales, and 40 l/m^2 for Scotland. Losses in the generation of domestic hot water maybe summarised as draw-off pipe losses, storage losses, losses in primary circulation pipes, and those encountered during the heating up by the hot water generator. For a realistic plant simulation, these losses should be taken into account.

2.3.3 *Solar Water Heating Systems*

There are three types of solar water heaters: thermosyphon, built-in storage, and forced circulation.

Thermosyphon solar water heater

A sketch of a thermosyphon water heating system is shown in Fig. 2.12. As the temperature of the water in the collector rises, its density decreases. A circulation is thus established which enables the tank water to be progressively heated. The temperature of water at any point within the circuit determines the corresponding density. If this density variation is plotted against the height contour of the circuit the magnitude of the driving force for water circulation is obtained. Owing to the low flow rates encountered within such systems, the driving force is balanced against

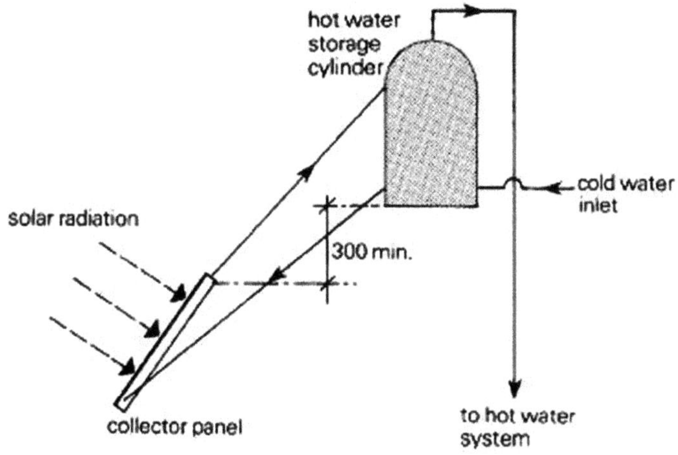

Fig. 2.12 Sketch of solar thermo-syphon heater

the friction offered by the fluid circuit. The flow rates can thus be estimated. To avoid flow reversals, it is recommended that the bottom of the tank should be at least 300 mm above the collector top. This requirement makes the system inconvenient, particularly if the collectors are to be roof mounted. Freezing may occur within the collectors during winter in a direct system. The solution is to employ a heat exchanger within the storage tank, with an anti-freeze solution added to water. This will however result in reduced system efficiencies.

Built-in-storage solar water heater

The construction of a built-in storage heater involves a rectangular box-like structure with the top face painted black and enclosed behind a single or double sheet of glass. The back surface and sides are properly insulated and the entire assembly is tilted at a suitable inclination. The built-in storage water heater possesses several advantages over the other types. First, during the day time, it has higher efficiency owing to the fact that primarily no heat losses occur during circulation of water. Second, in comparison with the thermosyphon and forced circulation systems, the intimate contact of water with the absorber plate results in better heat transfer. Built-in-storage heaters are also compact in size and cheaper due to their simplicity of construction. Based on extensive measurements of a number of designs [27] has reported the effect of storage volume/collector area ratio, number of glazing, and mode of operation on the performance of built-in-storage heaters. Kreider and Kreith [28] have presented design details of a freeze-tolerant, built-in-storage solar water heater with embedded heat exchanger coil through which a refrigerant is used to extract heat.

Forced circulation solar water heater

Figure 2.13 shows the schematic of a forced circulation water heating system. In this design, the tedium of placing the solar water tank above the collectors is removed. The water in Tank A is the feed water for tank B with a built-in auxiliary heater. In a more compact version of forced circulation system, the two tanks A and B are combined in a single unit. However, this reduces the overall efficiency of the system as the maximum exploitation of solar energy is inhibited. The overall efficiency of solar thermal collection devices is much higher than their PV counterparts and in this respect, a figure of 30–40% has been quoted by Courtney [29] who has provided thermal and economic analysis for solar water heating. The latter part of Courtney's study is now dated information. However, there has not been any major breakthrough in the thermal performance of flat-plat collector technology. Hence, the aforementioned part of the above work is still valid. It has been shown that south facing 4 m^2 collector system in the south-east of the UK would supply 1400 kWh of useful heat annually. For a temperature boost from 10 to 55 °C, it was shown that the solar fraction would be 40%.

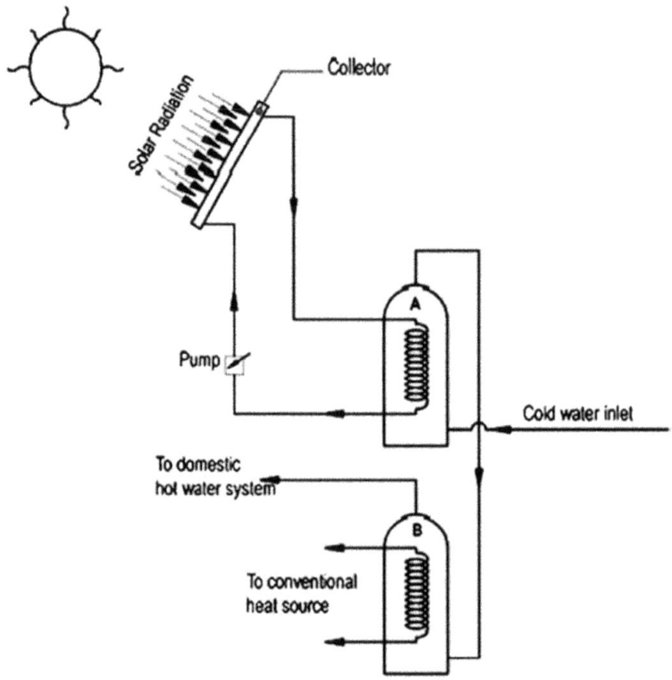

Fig. 2.13 Forced circulation solar water heater

Solar Thermal Collectors

By early 2010 the total energy generation of installed solar thermal collectors amounted to 185 GWth, of which China, Germany, Turkey, and India accounted for 80.3%, 3.1%, 1.8%, and 1.1%, respectively. The remaining 13.7% was accounted for other 40 plus countries including the USA, Mexico, India, Brazil, Thailand, South Korea, Israel, Cyprus, Ethiopia, Kenya, South Africa, Tunisia, and Zimbabwe [30].

Flat-plate hot water collectors

Figure 2.14 shows the constructional details of flat-plate collectors respectively using single- and double-glazing designs. A typical design involves a selectively coated copper absorber (flat-plate). The selective coating ensures maximum absorption of incoming short-wave solar radiation and low emission of thermal radiation from the plate. A serpentine shaped tube, or risers and header tubes carrying water (or water-based freeze-tolerant solution) transport heat from the absorber plate to the storage tank. The former design, owing to minimal connections and hence its lesser susceptibility to leaks is well suited for forced circulation systems. The latter design is, however, the right choice for thermosyphon systems as it ensures minimal pressure drop due to friction. An insulated collector housing that reduces the heat loss to a minimum surrounds the absorber. Typically, the collector housing consists of an

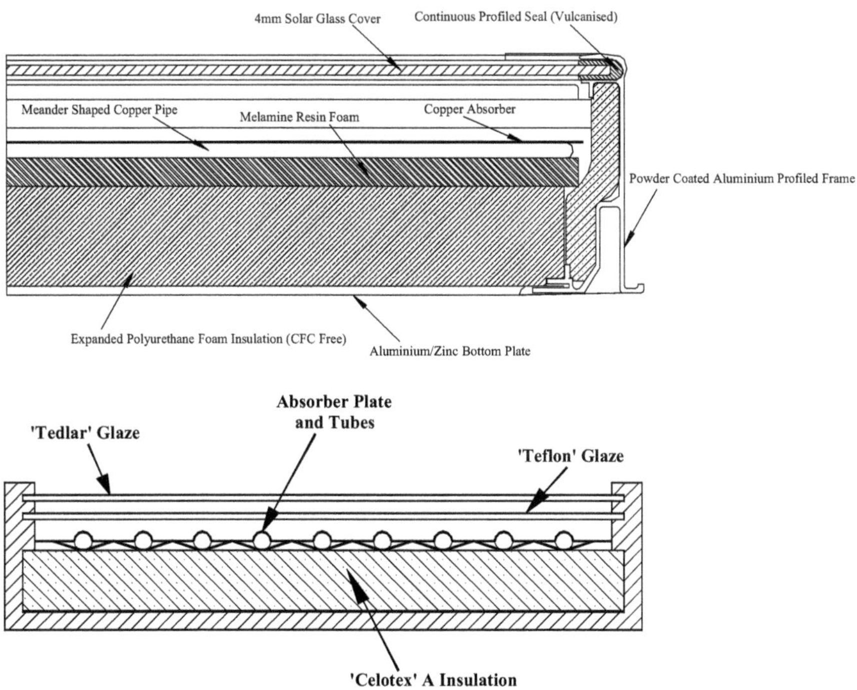

Fig. 2.14 Solar hot water collector designs

aluminium frame in which the glazing is secured by means of a profiled seal. Collector fittings are so designed that they may easily be connected in series, parallel, or series–parallel arrangement.

The heat loss coefficient from any given collector determines its thermal efficiency. This in turn is obtained from the heat loss from the collector top (UT) and the overall-loss coefficient (UL), i.e. the sum of heat loss from the top, sides, and bottom of the collector housing. For one particular, commercially available single-glazed collectors with selectively coated absorber UT and UL are respectively 3.37 and 3.4 W/m^2-K [31].

Evacuated-tube hot water collectors

Figure 2.15 shows the constructional details of evacuated-tube collectors respectively based on coaxial-tube and heat pipe designs. In the former design, a coaxial heat exchange pipe carrying a counter-flowing fluid is embedded within the absorber. The heat exchange pipe then feeds into a header. To enable maximum exploitation of solar energy each vacuum tube is pivot mounted, enabling optimal orientation. The

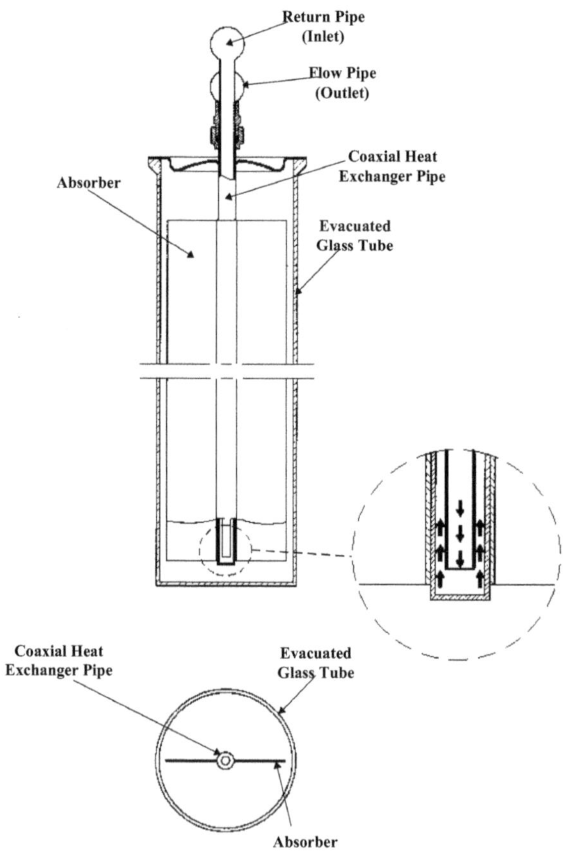

Fig. 2.15 Evacuated-tube collector

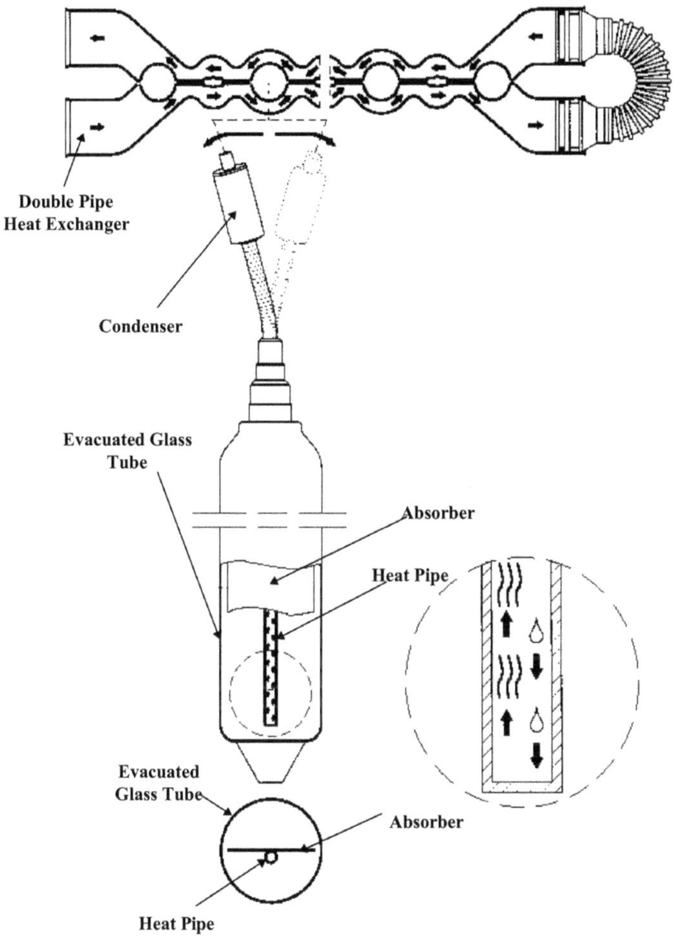

Fig. 2.16 A heat pipe solar collector

vacuum in the glass tubes ensures minimal convective heat loss while the selective coating helps suppress radiative losses. The top-loss and overall-loss coefficients for the collector presently under discussion are quoted as 1.37 and 1.375 W/m^2-K [31].

The design of the collector shown in Fig. 2.16 is based on the principle of heat pipe. Basically, a heat pipe is a closed container that employs an evaporating-condensing cycle. Transfer of heat from the absorber plate is via an efficient and fast heat conductor bearing a low heat capacity. The heat pipe functions as a thermal diode. The heat pipe accepts heat from an external source, uses this heat to evaporate the liquid (latent heat), and then releases latent heat by reverse transformation (condensation) to the heat sink. This process is repeated continuously by a return feed mechanism of the condensed fluid back to the heat zone. Typically, evacuated heat pipe solar collectors use a sealed steel pipe, containing alcohol. The pipe is then attached to a black copper fin that fills the tube (absorber plate). Protruding from the top of each tube is a metal tip attached to the sealed pipe (condenser). These tubes feed into a heat exchanger manifold. Under solar flux, the alcohol is heated and hot vapour rises to the top of the pipe. Water, or glycol, flows through the manifold and picks up the heat from the tubes. The heated liquid circulates through another heat exchanger and gives off its heat to water that is stored in a solar storage tank. The maximum operating temperature of a heat pipe is the critical temperature of the heat transfer medium. Since no evaporation/condensation above the critical temperature is possible, the thermodynamic cycle interrupts when the temperature of the evaporator exceeds the critical temperature. The top-loss and overall-loss coefficients for the above collector are quoted as 1.34 and 1.346 W/m^2-K.

Solar water heating economics

Lack of volume sales has maintained the price of flat-plate solar collectors at a relatively high level. The amount of heat energy captured per square metre of collector surface area varies with design and location but typically can range from 300–800 kWh/m^2/annum. Depending on water use behaviour and number of people in a household, solar water heaters can give long payback periods of around 5 to 10 years or longer. Within the UK, current installation costs for the average family home would amount to between £2000 and £3000 for a flat plate and between £3000 and £5000 for an evacuated-tube installation.

Based on current German cost data, system costs for a domestic hot water system with 6 m^2 collector area can be up to €1000 per m^2 or €6000 per system. The cost of a combi-system with 15 m^2 collector area at €900 per m^2 can reach €13,500 per system.

The investment costs for these types of systems (domestic hot water, combi-, thermosyphon) are assumed to range from €300 to €1000/m^2, with an average of €630/m^2. By 2030 system costs could decline by 35–50% to an average of €400/m^2, based on a global installed capacity of 200 GWth (investment cost being in the range of €150–€600/m^2). Assumed energy yields range from 250 kWh/m^2/year for a combi-system with a low solar fraction in, for example, central or northern Europe, up to 800 kWh/m^2/yr for a domestic hot water system with high solar fraction in southern Europe. The worldwide average yield is assumed to be 450 kWh/m^2/year based on regions with relatively favourable solar gains. By 2030, energy yield could increase by 11–20% to an average of 500 kWh/m^2/annum.

2.4 Solar Lighting

A high density of urban structures can often lead to the loss of daylight amenity. Improper design of large glazed facades may also cause the problem of glare. One innovative solution that has emerged in recent years is to capture daylight via a mirrored light-guide or lightpipe and then direct it to those areas of buildings that are starved of daylight. Monodraught, an English company specialising in building

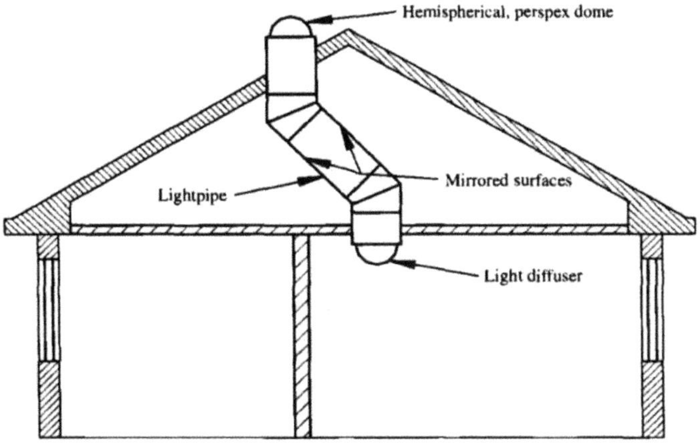

Fig. 2.17 Schematic of a light-pipe system

services products, has patented a 'Sun-pipe' system which can pipe daylight to internal spaces with the added advantage that excessive solar gains are avoided. The mirrored light transmission system is crudely analogous to the fibre optic cable, the difference being that in the latter device the transmission efficiency is much higher. A schematic of the Monodraught system is shown in Fig. 2.17.

Daylight is gathered via a polycarbonate dome at roof level and then transmitted downwards to interior spaces within buildings. The internal surface of the sun-pipe is coated with a highly reflective mirror finish material (typically with reflectance in excess of 0.95) which helps in achieving a reasonable illuminance indoors when daylight is introduced via a light diffuser. The light-reflecting tube is adaptable to incorporate any bends around building structural components. Measurements on sun-pipe systems were undertaken at Edinburgh Napier, Liverpool, and Nottingham Universities. These tests have indicated that under hazy sunshine conditions a lighting level of 900 lx was recorded internally at approximately 2 m below the outlet light diffuser. In typical installations sunpipes of up to 6 m lengths have been used, some with four pipe bends.

The building types with light-guides have ranged from insurance companies to health clinics and from residential to university buildings. A further innovation in the development of the sun-pipe system is its integration with electric lighting within the lightpipe so that the device continues to function as a light source even during the hours of darkness. In another innovation, the sun-pipe has been combined with a natural ventilation system that draws air from the rooftop opening to provide cooling for the building fabric.

The broad conclusions that were drawn were that lightpipes can be effective devices for introducing daylight into buildings, provided an excessive number of bends or long pipe lengths are avoided. Use of as large a pipe diameter as possible was also recommended. References [32–34] will present computational methods for design and performance assessment of light-guides.

The performance of a lightpipe is governed by a number of factors. These may be enumerated as:

The length of the pipe, the number of bends, the angle of each bend, the reflectance of the lightpipe (aluminium or total internal reflection film), the slope of the outdoor roof on which the lightpipe dome is situated, the location of the installation.

The interplay of these factors is available in References [32, 33] which deal with the modelling of such systems.

2.5 Use of Solar and Wind Energy: Challenges and Opportunities

As shown in Chap. 1 there is a plentiful supply of solar and wind energy that is available on this planet. Decreasing cost of renewables has now reached the stage that they are directly competing with fossil fuel and nuclear energy. However, there are many challenges that still need to be met such as low capacity factors and temporal nature of renewables. To address these the following sections present a few solutions.

2.5.1 Simultaneous Generation of Solar PV Over Wide-Spread Locations

Solar and wind energy are highly spatial and temporal in their nature. The spatial aspect may be handled by exploitation of more wind than solar energy at higher latitudes and vice-versa for mid-latitudes around the equatorial belt. However, as shown in Sects. 2.2 and 2.6 even at northerly latitudes solar energy is proving to be economical due to the price breakthrough of PV modules and an ongoing effort to increase module efficiency.

One way to solve the temporal issue is to install wind and solar power plants at diverse locations and in this respect Figs. 2.18 and 2.19 present some interesting information. The data were obtained from three UK locations that are separated widely. Aberdeen is in North-east Scotland, likewise, Finningley is on the east coast of England at a much lower latitude when compared to Aberdeen, but much more northerly when compared to Camborne. Refer to Fig. 2.18. The plots demonstrate the spatial aspect of solar energy—with Aberdeen delivering a high yield when compared to the other two locations which experienced a part-overcast and overcast regime.

One way to analyse the stability of a grid-connected PV-based energy delivery system is to obtain the coefficient of variation for the three individual locations and the combined resource and that is shown in Fig. 2.19. Note that the coefficient of variation (CoV) is the ratio of the standard deviation of the daily receipt of irradiation and its average, for each month. As the CoV is dimensionless it enables a direct comparison between locations and seasons. Note that, as shown in Fig. 2.19, the CoV for the

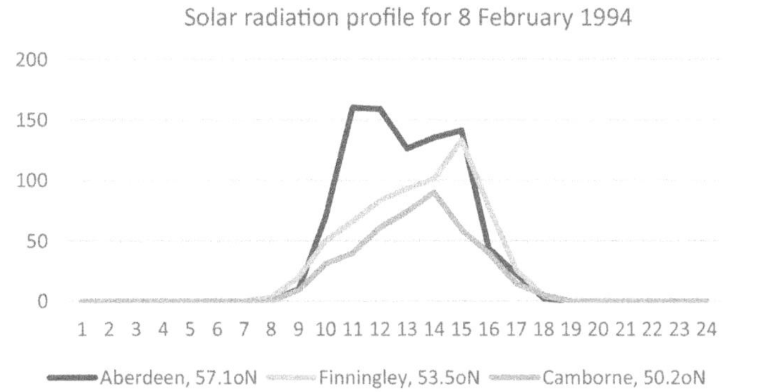

(a) Hourly solar radiation profile for three widely separated UK locations on a winter day.

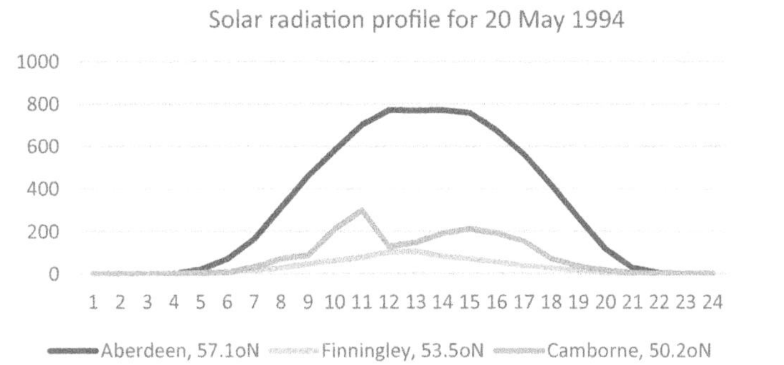

(b) Hourly solar radiation profile for three widely separated UK locations on a spring day.

Fig. 2.18 Hourly variation of solar irradiation at three diverse UK locations

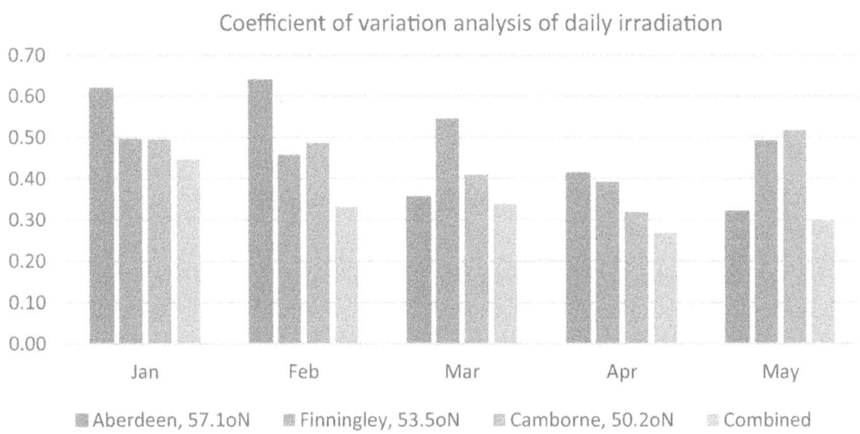

Fig. 2.19 Daily variation of irradiation analysed via coefficient of variation

Table 2.15 The complementarity of the wind and solar energy resource, relative units

Latitude, degree	Wind	Solar	Totals
10	0.35	0.5	0.85
33	0.425	0.425	0.85
50	0.475	0.25	0.725

combined resource from the three locations can be as low as half of the value for individual locations. By an extension of this argument if a very large number of locations were grid-tied that would bring down the variability of the resource very significantly.

2.5.2 *The Complementarity of the Solar/Wind Resource*

Refer to Fig. 1.10 that showed the latitudinal variation of the wind and solar energy resources. Let us take on board the case of inland locations. From the data presented in Fig. 1.10, we are now able to construct Table 2.15. The totals of wind and solar energy show near independence of latitude with solar being the more abundant resource near equator and wind dominating at higher latitudes. If wind and solar are of equal cost per kWh generated it would make more economic sense for wind to be used at higher latitudes and vice-versa. However, we find that even at higher latitudes solar is providing good competition to wind as the cost of the former is dropping much more rapidly, e.g. in June 2019 the UK, a high latitude country, had exceeded 13.3 GW of installed solar capacity with a total of nearly a million installations. The total UK wind capacity in the same year had exceeded 24 GW with more than 11,000 wind turbines installed.

The IEA 'Stated Policies Scenario' discussed in Sect. 2.2 also states that solar will become the largest source of installed capacity around the year 2035.

2.5.3 *Transmission of Electricity Over Long Distances*

One of the key technological developments that have to take place for reliable exploitation of renewables on a large scale will be transmission of electricity over long distances. With that capability, it will be possible to use solar power conversion in desert regions and its use in urbanised locations. Likewise, the offshore wind resource from high latitude locations with capacity factors exceeding a 40% mark may be used with confidence. Shown below are some examples of long-distance electricity transmission. There is an element of evolution taking place with longer distances involved and an affordable price tag. Note that most of the material present in this section is based on material presented in Power Engineering International journal [35].

(i) Vastervik-Ygne, Sweden (98 km, 1954, 20 MW)
VDC converter technology is based on the use of line-commutated or phase-commutated converters. In 1954, the Swedish company ABB, utilised mercury arc valves to construct the world's first commercial HVDC link between Västervik, on the east coast of Sweden, and Ygne, on the island of Gotland in the Baltic Sea. The original Gotland link could transfer 20 MW over a 98-km long submarine cable with a voltage of 100 kV. In 1970, the capacity was increased to 30 MW at a voltage of 150 kV through the addition of thyristor technology.

(ii) Belo Monte-Rio de Janeiro, Brazil (2543 km, 2017–2019, 4 GW)
The 2543 km Belo Monte-Rio de Janeiro transmission in Brazil is an 800 kV ultra-high-voltage direct current (UHVDC) line that transmits electricity from the 11.2 GW Belo Monte hydroelectric power plant located in Para to Rio de Janeiro, Brazil. Construction of this transmission line started in September 2017 and completed in April 2019. The overhead transmission line, which has transmission towers 105 m and taller, crosses 80 cities along its route from the Amazon to Brazil's southeast coast. The line has two power converter stations and is capable of transmitting 4 GW of electricity. The Belo Monte- Rio de Janeiro transmission line was built by State Grid Corporation of China. It was the second 800 kV UHVDC line to be built after the Belo Monte UHVDC line, which was completed at the end of 2017.

(iii) Rio Madeira, Brazil (2385 km, 2013, 7.1 GW)
The Rio Madeira transmission link in Brazil is a 600 kV high-voltage direct current (HVDC) bipolar line commissioned in November 2013. It is capable of transmitting 7.1 GW of electricity from Santo Antônio and Jirau hydropower plants on the banks of the Madeira River in north-west Brazil to major load centres in south-eastern Brazil. A consortium comprised of three major Brazilian energy providers developed this HVDC transmission line which was constructed in 24 months by Interligação Elétrica do Madeira. ABB supplied power equipment for three HVDC stations. Alstom supplied two HVDC bi-pole converter stations and four HVDC power converter transformers for the project.

(iv) Belo Monte-Estreito transmission line, Brazil (2092 km, 2017, 4 GW).
The Belo Monte-Estreito transmission line was the first 800 kV UHVDC transmission line to be constructed for delivering electricity from the Belo Monte hydroelectric power station in northern Brazil to south-east Brazil.

Completed in 2017, the 4 GW power capacity, 2092 km-long UHVDC line originates from Xingu in Para and terminates at Estreito in Minas Gerais.

(v) Jinping-Sunan transmission line, China (2090 km, 2012, 7.2 GW, 3.5 billion USD).

Completed in 2012 at a cost of 3.5 billion USD, the Jinping-Sunan electricity transmission link in China is an 800 kV UHVDC transmission line with a carrying capacity of 7.2 GW.

The line passes through eight Chinese provinces to transmit power generated from Guandi, Jinping, and Sichuan hydroelectric plants located on the Yalong River in central-western Sichuan province to the industrialised coastal area of Jiangsu province in eastern China. The AC voltage at both ends of the line is 525 kV.

(vi) Xiangjiaba-Shanghai transmission line, China (1980 km, 2010, 7.2 GW).
The world's first UHVDC transmission line, the Xiangjiaba-Shanghai transmission line, with an overhead length of 1980 km, is an 800 kV, 7.2 GW entity and has been in operation since 2010.

The Xiangjiaba-Shanghai link transmits power from the Xiangjiaba hydropower plant located in south-west China to the country's major industrial and commercial hub, Shanghai. The link comprises a single overhead line and 28 high/ultra-high-voltage converter transformers. The AC voltage at both ends of the line is 525 kV. Once again, ABB provided the overall system design, in addition to supplying the key components such as high and ultra-high-voltage power converter transformers, thyristor valves, DC and AC switchyard equipment, and the HVDC control system.

(vii) Inga-Kolwezi transmission line, Congo (1700 km, 1982, 560 MW).
Congo's 1700 km-long Inga-Kolwezi transmission project is a 500 kV line with a rated capacity of 560 MW. It is owned and operated by Democratic Republic of Congo's (DRC) national electricity utility, Société Nationale d'Electricité. The Inga-Kolwezi HVDC link carries power from the Inga Falls hydropower station on the Congo River to the copper mining district of Katanga in south-eastern Congo. The AC voltage at both ends of the line is 220 kV.

(viii) Talcher-Kolar transmission link, India (1450 km, 2003, 2.5 GW).
The 500 kV HVDC transmission line Talcher-Kolar is also known as the East-South transmission link. It has a rated capacity of 2.5 GW. The link transmits electricity from the Talcher power generation centre in the East Indian state of Orissa to Kolar near Bangalore. The bipolar transmission original line had the capacity to transmit 2 GW of power but upgraded to 2.5 GW in 2007. Siemens built the converter stations for the Talcher-Kolar HVDC link [36].

(ix) North-East Agra UHVDC link, India (1728 km, 2015, 8 GW, 1.63 billion USD).
Power Grid Corporation of India built India's first UHVDC, 800 kV North-East Agra transmission line which is capable of transmitting 8 GW. The project was delivered by ABB. It is the longest power transmission line in India. The total cost of this project was 1.63 billion USD [37].

In summary, if we examine the above two projects for which the costs data are available, i.e. the 2090 km, 7.2 GW Jinping-Sunan line in China and the 1728 km,

8 GW North-East Agra UHVDC Indian link the cost indices respectively come out as 0.13- and 0.07 US cents/kWh/thousand km distance travelled. The above calculation is based on the assumption of an 80% capacity factor and a 20-year project lifespan. Note that the Chinese project got completed in year 2012 and the Indian one some three years later which also shows the declining project cost.

2.5.4 Hydrogen Production and Transmission

It was shown in Sect. 5.1 that a UK-wide PV grid would substantially reduce the day to day fluctuations in PV supply should a single location be considered in isolation. However, even with the combined output of the high latitude locations, such as those of northern Europe, New Zealand, and southern Chile, the resource availability is fairly low during their winter months. It is presently hypothesised that by the year 2040, a serious effort would be made worldwide to make a switch of electricity production from fossil fuel to non-polluting sources of energy, such as PV solar. In such a scenario, only a small fraction of 'neighbouring' arid/semi-arid regions, clad with PV, would be required to supply the large quantities of pollution free energy. One of the two likely energy vectors would be hydrogen, produced by electrolysis of water and then distributed via pipelines with staged compression. It is proposed that hydrogen is produced using PV generated electricity within arid/semi-arid regions with proximity to the sea.

A figure of 4.2–4.6 kWh/m^3 of hydrogen at normal temperature and pressure is quoted for the electrolysis electrical energy requirement [38]. In the future, the latter production efficiency is predicted to improve to 3 kWh/m^3 by the year 2050 [39]. Presently, closer attention is paid towards the feasibility of a supply of hydrogen from Saudi Arabia to partially meet the year 2040 electrical demand for OECD-Europe. It is assumed that solar and wind (offshore and onshore) would be responsible for a significant share of the national energy supplies. Further, it is assumed that a proportion of the deficit energy would be acquired via PV solar electricity, produced in respective 'neighbouring' arid/semi-arid regions. For example, for Europe, the provision of the latter energy supply via hydrogen carrying pipelines from Southern Spain and Saudi Arabia may be considered.

The geography of Saudi Arabia is such that a narrow coastal plain on the Red Sea rims a mountain range that spans the length of the western coastline. These mountains gradually rise in elevation from north to south. East of these mountains is a massive plateau that slopes gently down towards the Arabian Gulf. Part of this plateau is covered by the world's largest continuous sand desert, the Rub Al-Khali, or Empty Quarter. This name is quite befitting, owing to the fact that save a few small hamlets, there are no major conurbations in the vast expanse of desert which has the following coordinates: 16 N, 45 E; 21 N, 45E; 25 N, 50 E; and 23 N, 55 E. The total area of Saudi Arabia is over two and a quarter million square kilometres and that of the Empty Quarter is 425,500 km^2. The latter region, thus, occupies 19% of the Saudi land mass. On the Eastern flank, the Empty Quarter snuggles with the Arabian

Gulf. It, therefore, has a good potential to be used for large-scale exploitation of solar energy with a stable supply of large-scale PV electricity and access to seawater for electrolytic production of hydrogen.

For a sustainable energy supply to Western Europe, a suitably designed compression plant will be able to supply hydrogen gas from the eastern most part of the Empty Quarter to Adana in the southern part of Turkey, an approximate distance of 2500 km. An identical booster station would then be able to transport the gas to southern Germany, which is a further 2300 km away. Note that the PV clad area of 2800 km^2, representing 0.6% of the area of the Empty Quarter, would be enough to supply a significant part of the OECD energy requirements.

Presently, a modular approach is proposed for implementation of PV solar generation in arid/semi-arid regions and transmission using hydrogen as the energy vector. The CIBSE handbook [40] provides guidance on the design of pipes for transmission of compressed gases. Assuming a three stage compression with a delivery pressure of 6 bar and minimum pressure in the pipeline of 2 bar, it may be shown that each 2 m diameter steel pipeline would be able to carry 11.5 kg/s of hydrogen [41].

Assuming a fuel cell efficiency of 78%, the energy carrying capacity of each pipeline would be 290 kWh/s which translates to an annual capacity of 9 TWh. Because of the low density of hydrogen, compared to natural gas, staged pumping would be required only after a distance of over 3000 km. This calculation compares favourably with the 300–350 km natural gas interstation distance.

There is also precedence, for the above proposed design in terms of transport of fluids over long distances, e.g. the Trans–Siberian gas pipeline.

The cost of PV solar energy generation and transmission may be broken down into the following stages: PV array installation and maintenance, electrical energy transmission from PV station to the electrolysis unit, electrolysis of water into hydrogen and oxygen, compression of hydrogen, costs associated with hydrogen transmission pipeline laying and maintenance, booster compressor stations and the fuel cell cost at the delivery end. Since solar cells produce direct current electricity, required by electrolysers for PV-hydrogen production, the extra cost of DC to AC conversion equipment is avoided. Hydrogen can either be produced at atmospheric pressure or then compressed to a high pressure to move it through the pipeline or alternatively, electrolysis can be performed at high pressure to avoid the compression step. However, atmospheric pressure hydrogen production involving compression work is the better approach, since the high pressure electrolysis approach involves more costly devices while yielding lesser efficiency.

Hydrogen, produced by electrolysis using day-time PV electricity supply, can be stored to level the hydrogen flow in the pipelines during the remaining hours of the day to maximise the use of costly long-distance pipelines. Hydrogen can be stored in natural gas wells to levelise the hydrogen input to the pipelines. Ogden et al. [42] have presented a detailed account of the relevant economics. Adjusted to the inflationary effects, the year 2025 cost breakdown for the constituent components of the hydrogen economy are: total cost (7), PV array with electrolysis (4.78), hydrogen compression (0.84), storage (0.25), hydrogen transmission (0.13) and fuel cells (1), all quoted figures being US cents/kWh.

2.5.5 *Decreasing Cost of Electric Battery Storage*

Bloomberg New Energy Finance reports that over the past ten years, lithium-ion batteries have fallen in cost by around 80%. A Lithium-ion battery pack that cost USD 1183/kWh in 2010 dropped in price to USD 156/kWh in 2019. Figure 2.20 shows the declining cost of lithium-ion batteries.

Some of the landmark projects that house large battery systems to support renewable energy are cited below:

(i) Since August 2020, Duke Energy of North Carolina has been operating a 9 MW/9 MWh lithium-ion Samsung battery system in the city of Asheville. The total cost of the project was less than $15 million. The system which includes frequency regulation and grid support services can support the electrical needs of a thousand homes for 4.5 h.

(ii) Duke Energy has plans to invest $600 million for 375 MW development of battery energy storage. The company has also been operating another 36 MW battery system to support its own Notrees Wind Facility in Texas. Another of its projects includes a 5 MW solar PV facility for the town of Woodfin in Buncombe County, North Carolina [43].

(iii) To date, the largest active battery storage project is LS Power's 250 MW Gateway project which is located in the East Otay Mesa community in San Diego County, California. The team behind this project was NEC that built the system with LG Chem batteries and SMA inverters. The storage capacity is designed to be between 1- and 1.5 GWh. Furthermore, LS Power has in its pipeline the 200 MW Diablo Energy Storage facility in Pittsburg, California, the 125 MW LeConte Energy Storage facility in Calexico, California, and the

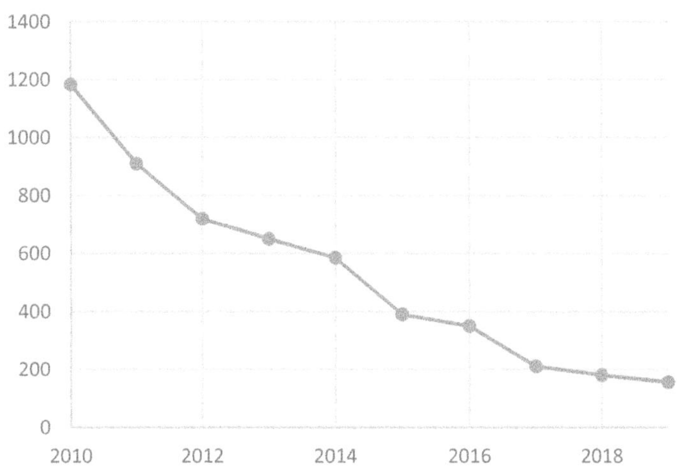

Fig. 2.20 Declining cost of Lithium-ion batteries: Cost units are relative. *Source* Bloomberg NEF, May 2020

316 MW Ravenswood energy storage project under development in Queens, New York.

(iv) In year 2020, Vistra Energy secured a permit to expand and construct at Monterey County, California energy storage system which has 1500 MW power and 6000 MWh energy storage rating.

(v) Tesla is building a 182.5 MW/730 MWh battery featuring a 256 T Megapack battery, which will be fully complete in the second quarter of 2021.

(vi) There is also the upcoming 409 MW Florida Power and Light Manatee project, set to be the largest battery installation on the U.S. east coast. Note that items (iii)–(vi) were sourced from www.pv-magazine.com.

(vii) Tesla-operated Hornsdale Power Reserve in South Australia is another example, that is capable of storing 150 MW/193.5 MWh [44].

2.5.6 The Role of Electric Vehicles

According to Wards Intelligence, a survey carried out in year 2016 indicated that, globally, there were over 1.32 billion vehicles. That figure is expected to rise to 2.8 billion by 2036. In the US alone there were over 268 million vehicles in the year 2016 and that population is increasing by 17 million per annum. China seems to have the most cars though with a count of 301 million in April 2017 and increasing at a rate of 28 million/year. The large automobile population can be made to be a useful storage facility for renewable energy provided it is switched to Battery Electric Vehicles (BEV) design. That point will be discussed later on in this chapter.

The runaway success of the automobile is due to its impressive performance—travels more than 500 km without refuelling the fuel tank and can attain speeds of over 160 kmph.

The global auto industry is a key sector of the economy for every major country in the world. The industry continues to grow, registering a 30% increase over 1995–2005. Building over 66 million vehicles a year (year 2005 statistics) requires the employment of about 9 million people directly in making the vehicles. That is over 5% of the world's total manufacturing employment. It is estimated that each direct automobile job supports at least five indirect jobs in the community, resulting in more than 50 million jobs. Figure 2.21 and Table 2.16 are referred here to strengthen the above argument [45].

One of the fastest growing sectors for electricity consumption is due to rapid emergence of hydrogen-powered and Battery Electric Vehicles (BEVs).

A considerable amount of R&D money has been spent by automobile manufacturers on the development of (a) hydrogen-powered vehicles that either use fuel cells or internal combustion engines, and (b) Battery Electric Vehicles. In March 2003, the Japanese Honda company introduced the first fuel cell car. The model FCX used polymer proton exchange membrane technology and develops 81 kW brake-power by the reaction of hydrogen (stored at 350 bar pressure) and atmospheric oxygen.

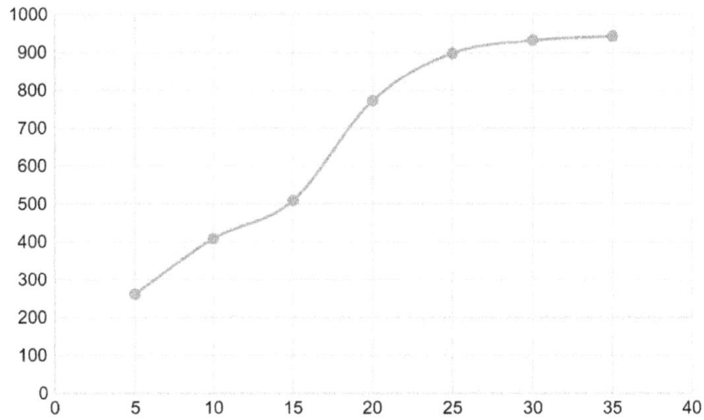

Fig. 2.21 Relationship between car ownership (vehicles/thousand persons: *y*-axis) and affluence (annual income, thousand USD/annum: *x*-axis)

Table 2.16 Pros and cons of an electric vehicle

Advantages	Disadvantages
Quiet and quick	Limited range
Home recharging	Long refuelling time
Cheaper to operate	Higher cost
No tailpipe emissions	Lack of consumer choice
Higher efficiencies	Lower energy density

Based on Berman, B., & Shenoi, R. (2016). Electric cars pros and cons [online]. PluginCars.com. Available at: http://www.plugincars.com/electric-cars-pros-and-cons-128637.html

Note that the Fuel Cell Vehicles (FCV) are also electricity powered as the required hydrogen will mainly be produced via electrolysis.

The performance of electric vehicles is developing all the time. American electric vehicle start-up Lucid Motors is in the process of manufacturing 'Air'—a model that is capable of a speed of 235 mph. The 'Air' EV model features a 75 kWh battery pack from Samsung and an electric motor producing 395 bhp (295 kW). The four-wheel-drive model gains a second motor for a total system output of 1000bhp (746 kW) and is expected to manage more than 400 miles per charge supported by a 130 kWh battery [46].

In terms of leading the electric vehicle revolution, Norway is way ahead of the rest of the countries of the world. The data emerging from Norway is quite spectacular in this respect: (i) The proportion of electric vehicles registered in September 2020 were 62% of the total. That proportion rises to 82% if Plug-in Hybrid Electric Vehicles (PHEV) are included. (ii) By year 2025 Norway expects all new passenger cars and light vans sold will be zero-emission vehicles. All new urban buses sold shall be zero emitters or use biogas. (iii) A third of the entire country's vehicle fleet will be pure electric. The latter proportion is expected to rise to 60% by the year 2030. (iv) In year 2045 all fossil fuel automobiles are expected to be phased out. (v) In year 2020 whereas an EV cost Euro 33,250 the price of a mid-sized, petrol car, including taxes, was Euro 34,000.

The 2018–19 versions of some common electric car models display a battery energy density that is 20–100% higher than their year 2012 counterparts. Furthermore, battery costs have decreased by more than 85% since 2010. For the next five years, automakers have announced plans to release another 200 new electric car models.

FCV sales have however faltered in the immediate past.

In the US, there was not much progress in 2019 with around 7500 units. Those sales figures were 90% more than those from a year earlier.

The bulk of the result comes from a single model—the Hyundai NEXO, which with sales of 4818 in South Korea overshadows Japanese FCVs like Toyota Mirai (2407 cars sold) and Honda Clarity Fuel Cell (349 sold) [47].

The BEVs, however, are enjoying good sales with around 2.2 million passenger plug-in car sales globally in 2019 which translates into an average of 2.5% market share [48].

Charge station network has also to be developed to meet the energy demand of growing population of BEVs. Within the UK alone charging points increased from a few hundred in 2011 to 32,200 by end of June 2020 at 11,600 locations [49].

2.6 Case Studies

2.6.1 Edinburgh Napier University Building-Integrated Photovoltaic (PV) Installation

In April 2005, Edinburgh Napier University installed a solar Building-Integrated Photovoltaic (BIPV) system, costing £155,000 comprising 32 rows of BP silicon solar panels, covering a total nominal area of 160 m^2. The complete system is connected to the university grid. Direct current (DC) power is produced from the BP-Solar monocrystalline panels, each of which produces 90 W of power at 22 V, which is then converted to a stable AC supply by four inverters. There are a total of 192 modules, connected in four strings, each string is connected to an individual inverter. There are two inverters of 6- and the other two of 2 kW each capacity. The 6 kW inverters get their feed from 72- and the 2 kW units from 24 modules. The highest value of the generated peak power to date is 13.7 kWp which occurs on a clear day in October. That depends on the sky clarity and the angle of the sun. The modules have an aspect that is 37° East of South and are inclined at 75° angle. Figure 2.22 presents a photograph of the facility under discussion.

The four Fronius inverters mentioned above are connected thus. Two large (IG60) inverters are connected to four strings of 18 modules wired in series; each of these

Fig. 2.22 Solar panels installed on the southeast face of Edinburgh Napier University—Merchiston campus

strings is wired in parallel with the others. Two small (IG20) inverters are connected to two strings of 12 modules wired in the same way as the IG60 inverters. Each of the two IG60 inverters receives from the PV panels a nominal power output of 4.6 kW at 20 A. Similarly, the two small IG20 inverters receive 2 kW nominal power at 7.8 A. Each of the IG60 and IG20 inverters consumes, respectively, 12- and 7 W of electric power during their operation and 0.15 W each during the night. The AC output from the PV inverters is fed to a sub-distribution board, which feeds into the building's main distribution board.

The fully instrumented PV facility recorded input (incident solar energy) and PV electrical energy output, at a frequency of 15 min.

PV output calculation: Table 2.17 provides the calculated and measured cumulative energy quantities for the PV facility under discussion. The differences between the calculated and the measured cumulative energy are due to seasonal variations that lead to differing frequencies of cloud cover and hence the colour of the sky. An overall efficiency of 11.5%, was obtained for the irradiation AC power conversion.

Incident solar energy: To measure the incident irradiation, two high quality pyranometers, CM11 and CM6b, manufactured by Kipp and Zonen of the Netherlands, were used.

Table 2.17 Cumulative energy assessment for the pv installation

Month	Calculated global tilted irradiation (MWh)	Measured global tilted irradiation (MWh)	PV AC output (MWh)	PV system conversion efficiency (%)
April	10.4	Not measured	1.6	Not calculated
May	13	Not measured	1.5	Not calculated
June	12.5	10.3	1.2	11.1
July	12.6	12	1.3	11.2
August	11.6	11.2	1.3	11.9
September	8.8	9.7	1.2	12.3
October	6	5.4	0.6	10.6
Total[a]	51.4	48.6	5.6	11.5

[a] (excluding April and May)

DC and AC Power output: Photovoltaic DC and AC outputs were measured via data-logging equipment installed within the inverter housing. This data-logging equipment permits storage of the data for the whole system over long periods (up to 6 weeks at 15 min frequency) and recorded data can be viewed on a computer screen using 'Fronius-IG Access' software. It records, averaged value of AC and DC voltage, current, and cumulative AC power with energy outputs for each inverter, at 15 min intervals.

Local meteorological Measurement: An hourly dataset covering a 27-year period containing horizontal global and diffuse irradiation based on the local Meteorological Office records at Turnhouse, Edinburgh airport was used to produce estimates of slope irradiation for the given PV aspect of 37° east of south and a tilt of 75° from horizontal. Figure 2.23 shows the long-term, hourly averaged (averaged for the entire period of 27 years) horizontal global and diffuse irradiation profile that is based on the above data set.

Analysis and Results: Fig. 2.24 presents performance of the Napier PV installation on a 15-min averaged-power basis.

The AC electrical output E can be obtained from the tilted global irradiance I_{TLT}, the cell area of the PV array $A = 125.5\ \text{m}^2$, and the overall efficiency η as,

$$E = I_{\text{TLT}} A \eta$$

Using above equation, an analysis may be undertaken to obtain the optimum PV module aspect and tilt. The chosen aspects for Edinburgh were east, southeast, south, south-west, and west. For each of the aspects, the following tilts were chosen: 0° (horizontal), 10°, 20°, 30°, 35°, 40°, 45°, 50°, 60°, 75°, and 90° (vertical). The result of the above analysis is presented in Fig. 2.25. It was found that for Edinburgh, the optimum aspect is south with the tilt range between 35° and 40°, producing the highest yield. The above yields are also compared with the Napier University installation with an aspect of 37° east of south and a tilt of 75°.

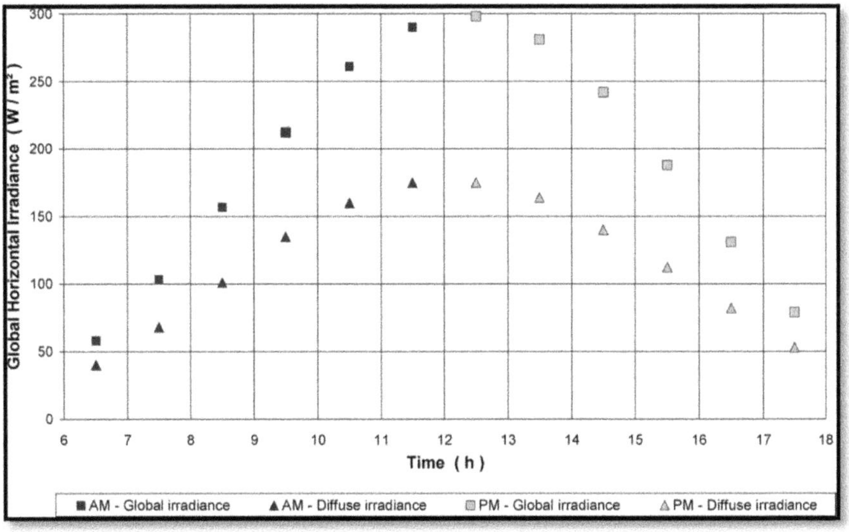

Fig. 2.23 Scatter plot of morning and afternoon, horizontal global and diffuse irradiance based on 27-year hourly Edinburgh irradiance data

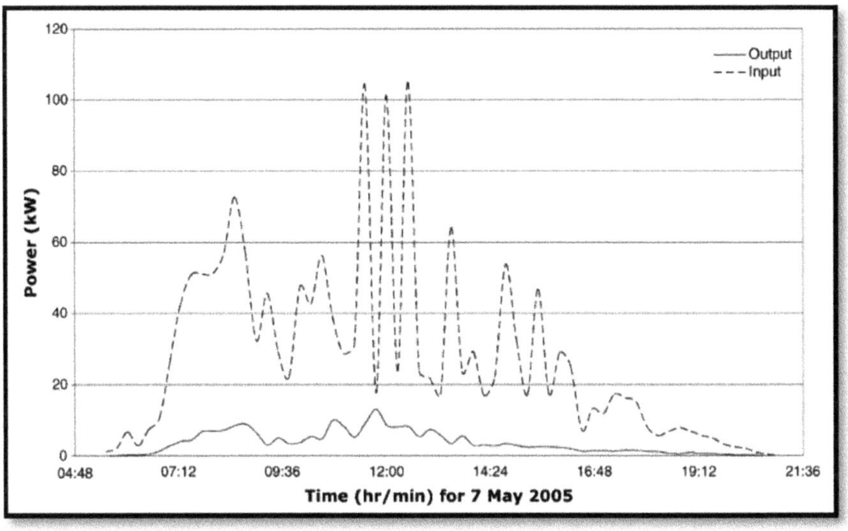

Fig. 2.24 PV incident power input compared with AC output for 7 May 2005

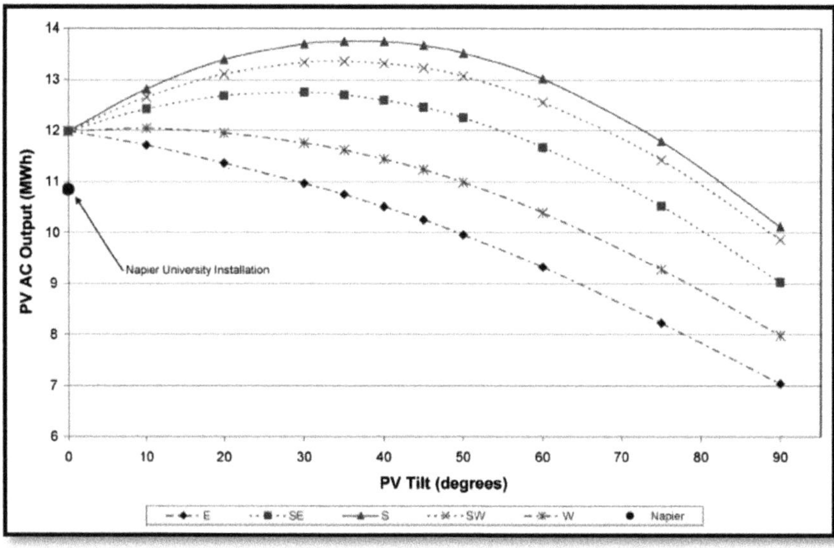

Fig. 2.25 Average annual PV AC output for chosen orientation and tilt

Note that the values for aspect and tilt were chosen due to architectural and planning restrictions. Hence, rather than generating 14 MWh_e AC-, the facility is generating 11 MWh_e AC output per annum.

Life cycle assessment (LCA): An LCA was performed to evaluate the embodied energy of the installation as well as the environmental impact of the installation. A list of materials that were used in constructing the given PV facility is provided in Tables 2.18 and 2.19. As a result, it was found that Edinburgh Napier University PV installation has 230 MWh_{th} and 5 MWh_e of embodied energy and this is equivalent to 14.5 metric tonnes of CO_2 emission.

Further, to calculate energy payback time (EPTB), Table 2.20 has been prepared.

Table 2.18 LCA for the PV installation: embodied thermal energy

Element	Materials	Mass (kg)	Embodied Energy (MWh_{th})	Co_2 released (kg)
Total spigots	Steel	200	3.3	440
Total vertical rails	Steel	2000	33	4400
Total tie brace	Steel	250	4.2	570
Total shuco rails	Steel	500	8.4	1100
Total PV module	Mixed	1500	180	3500
Cables	Copper	150	2.8	760
Total			230	11,000

Table 2.19 LCA for the PV installation: embodied electrical energy

Element	Materials	Mass (kg)	Embodied energy (MWh_e)	CO_2 released (kg)
Inverters, IG60 (2)	Mixed	40	2	1700
Inverters, IG20 (2)	Mixed	24	1	700
Operation			2	1100
Total	5	3500		

Table 2.20 Data used for LCA calculations

Elements	Global warming potential (g CO_2/kg)	Energy coefficient (MJ/kg)
Steel galvanised from ore	2200	60
PV module silicone m-c	2400	430
Inverters (2 IG60)	42,000	200
Inverters (2 IG20)	28,000	140
Cables	5200	70

The embodied energy payback time (EPTB) is defined as the ratio of embodied energy, converted to electrical energy, and the annual electrical energy production.

The EPTB for the Napier PV facility in the present study is estimated as 8 years. This payback period is compared with other similar installations in Table 2.21 [50, 51], which give results of the same order.

In addition, Fig. 2.26 shows the energy payback scenarios with respect to PV facade aspect and tilt variation.

Note that if the aspect and tilt were set to optimum, then the payback times would reduce quite sharply to just over 6 years. However, other practical considerations may hinder the adoption of optimum design.

Table 2.21 Energy payback

Installation	EPBT (years)	Incident solar irradiation (kWh/m^2/year)
Kannan et al	5.9	1635
Kato et al	15.5	1427
Present study	8.0	800

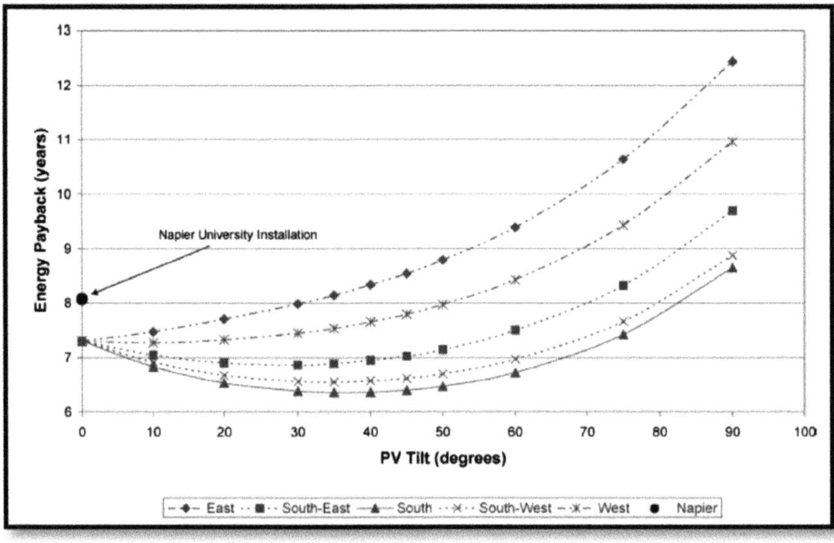

Fig. 2.26 Energy payback for chosen orientation and tilt

2.6.2 Solar PV House, Cardiff, Wales

PV system monitoring is not universally adopted by system owners which could be due to a number of reasons, including inaccessibility of the inverter display, i.e. located in loft spaces or garages, cost of wireless monitoring devices. The project under discussion was carried out at Cardiff University, Wales to evaluate the performance of a rooftop Solar PV system [52].

The PV system

By a precise architectural survey with a theodolite, it was found that the roof was positioned at 5° South of South-East and the roof inclination angle was 38° from horizontal. The size of PV system was 2.45 kWp. Ten frames mounted 245 Wp monocrystalline PV modules were connected in series to one frequency transformer string inverter. Maximum Power Point Tracking (MPPT) was utilised to enable the system to react quickly to rapid changes in insolation (incident solar radiation).

System monitoring

The inverter converts Direct Current (DC) generated from the PV array to Alternating Current (AC); while simultaneously logging power data that is sent in 10 min intervals via Bluetooth wireless communication to a netbook Sunny Explorer. A constant text file lists key system information for SMA Uploader. A Visual Basic software integrates with the PV output website Application Programming Interface (API).

The live data upload to the website occurs throughout the day using Window Scheduler. Email alerts can be setup if no updates are received, or if the average power generation drops below a defined value, over a specified time range. Mobile applications were used in the case study for Android and Windows environments for live remote PV data access.

The three software packages were used to model the system: PVSOL Epert v6.0, PV Watts, and PVGIS.

Results

A simulation package calculated energy generation based on MeteoSyn, a climate database for Cardiff latitude and longitude, averaged between 1961 and 1990. PV Watts (NREL) referenced an International Weather for Energy Calculation (IWEC) London climate (the closest available geographical location to Cardiff). The PVGIS-classic solar radiation database was referenced. The complete monthly, annual, and per cent error between actual and calculated energy (kWh) is described in Table 2.22.

The simulation software packages investigated gave annual energy generation figures which were lower than the actual energy generation from the system. PV Watts gave the lowest overall error of −4.6%, PVSOL with −5.6%, and PVGIS −7.4%. The annual yield of the case study was 903 kWh/kWp. The percentage errors between case study and simulated data, calculated on a monthly basis are shown in Table 2.23.

Table 2.22 Monitored and simulated energy generation (kWh)

Year 2012	System kWh	PVSOL kWh	PV watts kWh	PVGIS kWh
Jan	66.5	69.0	85	63.9
Feb	110.5	96.1	100	101
Mar	219.8	169.4	152	162
Apr	225.9	214.2	232	237
May	320.1	285.8	270	272
Jun	243.3	279.9	246	261
July	290.2	268.0	265	275
Aug	231.9	239.5	260	242
Sep	218.4	195.4	202	187
Oct	145.5	141.1	154	123
Nov	75.1	75.226	92	75.3
Dec	65	55.022	53	48.1
Total	2212.1	2088.8	2111	2047.3
Error		−5.6%	−4.6%	−7.4%

Table 2.23 Per cent error between case study and simulated monthly energy generation

Month	PVSOL	PVWatts	PVGIS
January	3	29	−3
February	−12	−10	−10
March	−22	−30	−25
April	3	3	3
May	−10	−15	−8
June	13	1	7
July	−8.5	−9	−8
August	3	11	4
September	−10	−7	−12
October	−2	8	−2
November	0	22	0
December	−15	−19	−29

2.6.3 Edinburgh College Solar PV Farm

A solar PV plant was constructed at Edinburgh College Midlothian Campus near Dalkeith in Edinburgh, UK. The plant was operational by the year 2012, occupies 2 hectares (5 acres), and has 2560 CSUN 245-60P polycrystalline modules. The modules were produced by China Sunergy. The peak power output of the plant is 627.2 kW. The electricity generation company Scottish and Southern Energy (SSE) who partnered with the project quoted the system cost for the entire project as £879,356, while the annual generation figure given is 568,611 kWh. From the above data, the capacity factor of the plant may be worked out as 10.3% and that concurs well with other installations in Central Scotland as shall be shown later on in this chapter.

A layout of the farm is shown in Fig. 2.27.

Plant specification

Nominal power output of each module is 245 W with a cell efficiency of 16.78%. Each module has 60 photovoltaic cells. Power One Aurora Trio inverters are used with a maximum power input of 20 kW each and nominal efficiency of 98%. Each inverter has two maximum power point tracking inputs, which ensure that the power conversion efficiency of the system is as high as possible, regardless of the amount of solar irradiation.

Experimental site

The complete evaluation of performance was made with the help of pyranometers, heat flux sensors, thermocouples, and data loggers. The complete setup is described in this section. Three pyranometers were installed between two modules (see Fig. 2.28).

As shown in Fig. 2.29, four flux sensors and two thermocouples were installed on

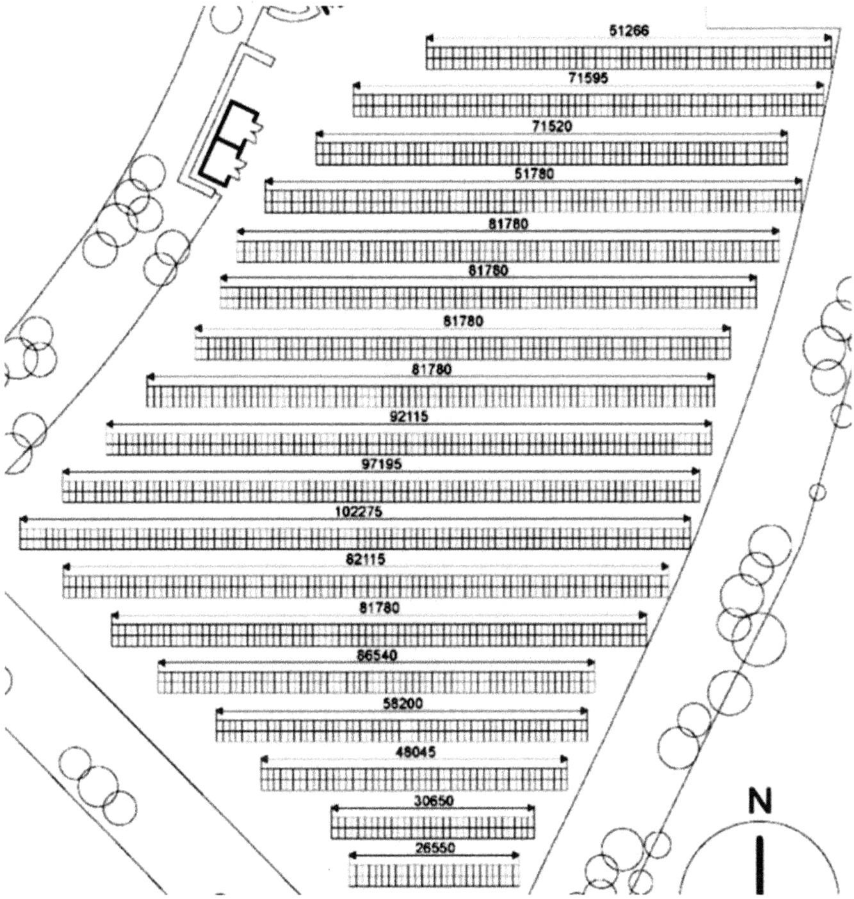

Fig. 2.27 Layout of Edinburgh solar farm

the back of the solar module.

Two, Grant 2020 series 'Squirrel' data loggers were used in this experiment. One data logger was used for heat flux sensors and the other for recording output from the thermocouples and pyranometers. They were kept in a metal storage box that was bound to the solar panel framing. To allow the sensors and wires to enter the box, a hole was drilled in the side of the box.

Results

Figure 2.30 shows an example of solar irradiation on a part-cloudy day. These readings were taken with the help of pyranometer 'p2' and 'p3'.

The ambient and cell temperature readings were recorded using thermocouple sensors denoted as 't1', 't2', 't3', and 't4'. The first two sensors measured the cell temperature from the back of panel and the other two were measured ambient (air)

Fig. 2.28 Pyranometer Setup. The location of pyranometers is marked in red boxes. Note that the pyranometer 'p1' measures horizontal- while 'p2' and 'p3' recorded slope irradiation

temperature. Figure 2.31 shows the resulting plot with the traces of 't1' and 't2' validating each other, likewise, traces from 't3' and 't4' outputs, once again, are superimposed.

Further information is provided in Jeffrey et al. [53].

2.6.4 Errol Solar PV Farm, Perthshire, Scotland

This farm was developed by Elgin Energy of the UK. A picture of the Errol solar PV farm is shown in Fig. 1.8. Some of the highlights of this solar PV farm are:

Location: Perthshire. Some 2.6 km west of Glenfarg and approximately 5 km northeast of Milnathort. The site is situated northeast of the Loch Leven basin and the M90 trunk road and is within an area of undulating topography and conical hill summits which mark the eastern extent of the Ochil Hills [54].

- Peak capacity: 13 MW
- Land area: 72 acres (29.14 ha, 2914 acres or 291,374 m^2)
- Modules used: 55,000 Canadian Solar 72-cell polycrystalline 315 W modules (CS6P-315P).

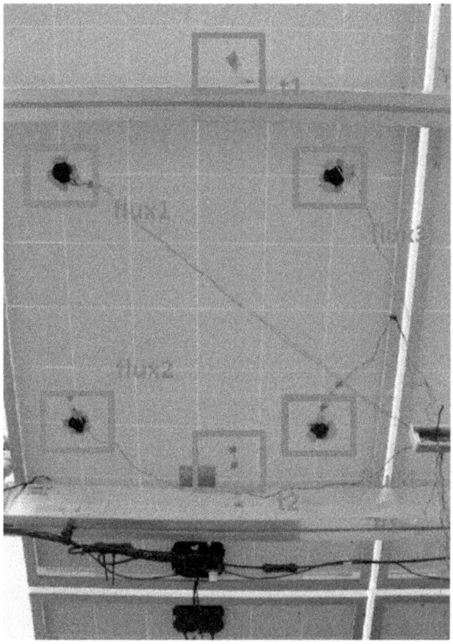

Fig. 2.29 Location of heat flux and temperature sensors. The red boxes indicate the position of heat flux and temperature sensors

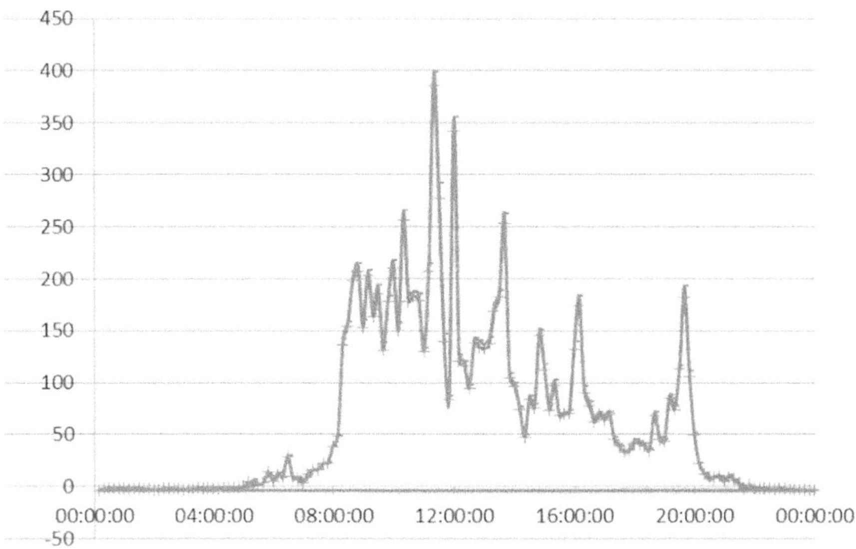

Fig. 2.30 Slope irradiation measurements (W/m^2). Note that the two traces recorded output from pyranometers, 'p2' and 'p3' are so close that they superimpose on each other, thus validating each other

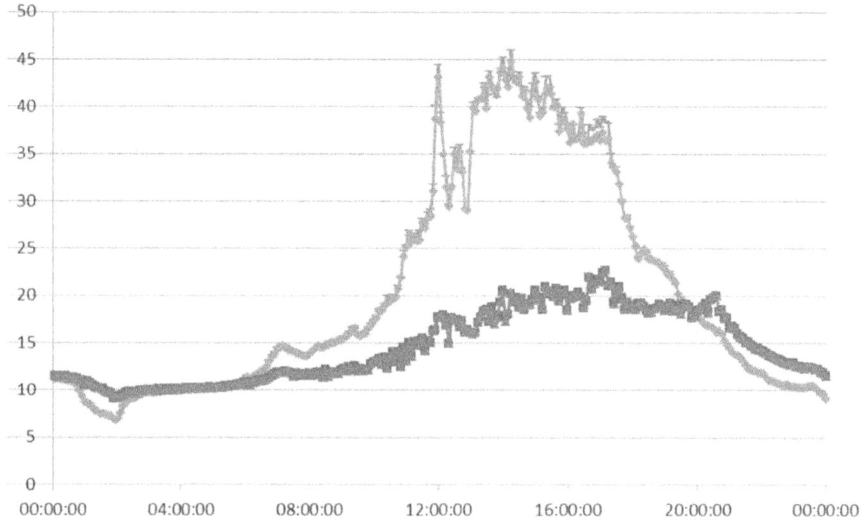

Fig. 2.31 Thermocouple temperature readings (°C)

Design and layout: The solar panels are arranged in rows, with each row feeding to an individual string. Each string is 4–6 m apart to prevent overshadowing and allow maintenance. Each string is screwed into the ground rather than concrete foundations to allow the land to return to agricultural use.

Each module is mounted on metal racks and at their highest point, they are 2.2 m high. All modules are angled at 25° and are facing the geographical south. One electrical substation supports the farm. The substation has dimensions of 8 × 4.95 × 4.95 m. The panels are mounted at around 0.8 m from the ground at the lowest edge, rising to around 2.2 m at the highest point. The panels measure 1.46 – 1.49 m × 0.99 m and are encased in an aluminium frame mounted on galvanised steel posts.

There are a total of 24 power inverters that collectively feed the electrical power to the national grid. The farm is monitored remotely. The site is monitored by 26 CCTVs mounted on 3 m high posts. Infra-red technology is also used.

Experimental research undertaken at Edinburgh Napier University under which a Renault-Zoe electric vehicle was driven for six years showed that on a typical day the city-wide journeys needed 4 kWh of battery charge. Hence, this farm can support over 15,000 electric vehicles.

2.6.5 Solar and Wind Energy Farm, Arbroath, Scotland

Solar Plant Performance: The 50 kWp of solar PV modules at North Mains of Cononsyth was installed in October 2011 with a Feed-in Tariff of 31.5 p/kWh. On average it produces 45,000 kWh per year. Due to inflation, the Feed-in Tariff was increased

Fig. 2.32 Cononsyth, Arbroath, Scotland 330 kWp wind turbine

from 31.5 p/kWh to 40 p/kWh currently earning £18,000/annum. Because the income of the export is so small the charge to export is almost as much as the export income. The total cost for installation was £150,000, giving a payback time of 9 years. The solar modules have an average capacity factor from year 2013 to year 2018 of 10.4% for an inclination angle of 15°. Year-on-year, the capacity factor has been steady.

Wind Turbine Performance: The 330 kWp wind turbine was installed in 2012 and with a FiT of 21.5 p/kWh, on average it produces 909 MWh annually (see Fig. 2.32). Due to inflation, the FiT has risen from 21.5 p/kWh to 23.93 p/kWh, currently earning £211,000 per year. On average the income from exported energy is £35,000/annum. The total installation cost of the turbine was £850,000 [55]. Indeed, by the year end of 2016, the turbine played itself back including interest having a payback time of 4.25 years.

The turbine performance is analysed by calculating its capacity factor for each of the six years. Figure 2.33 shows the capacity factor of the turbine from 2013 to 2018. The average capacity factor is 31.5% which is above average for onshore wind as per the guidance of the Department for Business, Energy and Industry 2018 report [56].

Figure 2.34 shows the performance of the wind turbine for year 2015, a windy year. The total generation was 1035 MWh with a capacity factor of 35.8%. For the rest of the UK in that year the capacity factor was 33.3%. The year 2016 was however a much quieter year with the turbine producing only 804 MWh with a capacity factor of 27.8% corresponding to the rest of the UK having a capacity factor of 23.6%.

In contrast, the Cononsyth solar PV plant, shown in Fig. 2.35, delivers a capacity factor of 10.4% compared to the UK average of 10.8%

Figure 2.36 provides a comparison of the Cononsyth wind turbine and solar PV facility. In this plot, both energy generation systems have been scaled to 100 kW for ease of comparison. Whereas the solar PV plant performs satisfactorily in summer the turbine underperforms. Likewise, in winter months the wind turbine consistently delivers. Thus, from an energy security point of view, a combined wind-solar facility would be desirable.

In examining the above data it should be noted that Scotland has a solar radiation income of 800–1000 kWh/m^2/annum, while the south of England has a solar radiation receipt of 1100–1300 kWh/m^2/annum.

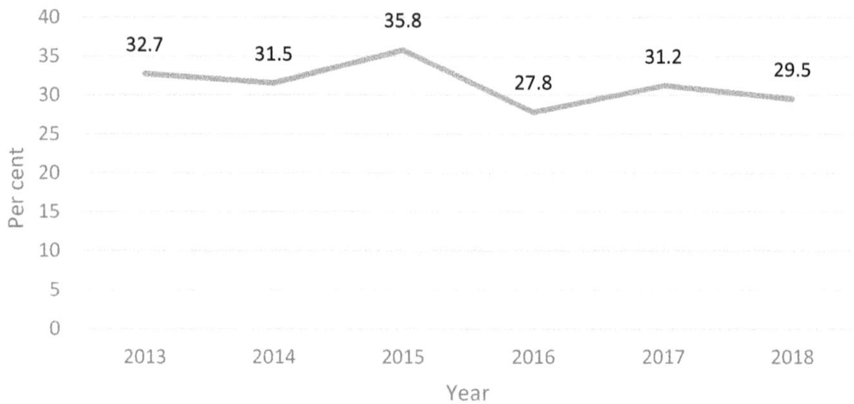

Fig. 2.33 Cononsyth wind turbine capacity factor for years 2013–18. Average capacity factor = 31.4%

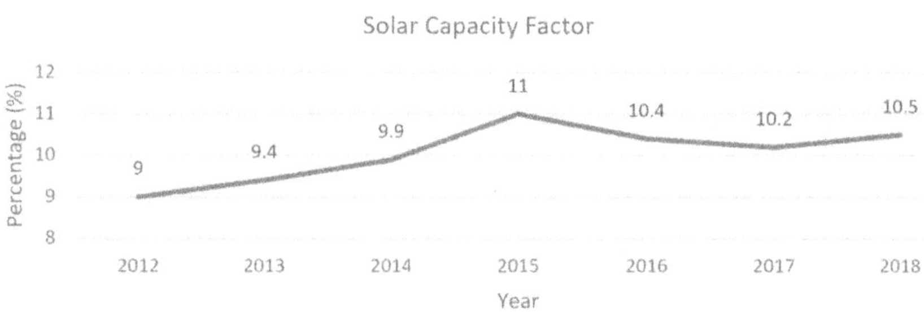

Fig. 2.34 Cononsyth solar PV plant capacity factor for years 2012–18

Fig. 2.35 Cononsyth, Arbroath, Scotland 350 kWp solar PV plant

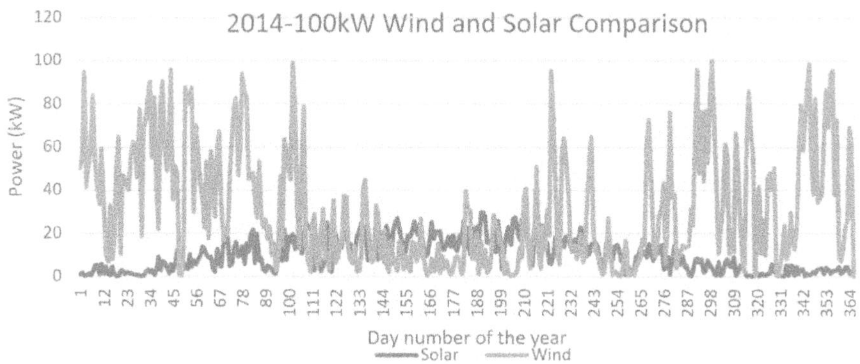

Fig. 2.36 Cononsyth wind turbine and solar PV facility generation performance comparison, both outputs scaled to 100 kW

2.6.6 Solar Water Heating for Textile Industry, Pakistan

Pakistan is amongst the most prominent cotton producing countries in the world. The demand for cotton has led to a well-established textile sector in Pakistan. A large proportion of its cotton products go into export. Being a successful candidate in international textile market Pakistani textile industry is continuously seeking modern and high-tech facilities to improve quality of its products. Textile industry has been the backbone of Pakistan's economy. It contributes more than 60% to the total export earnings of the country, accounts for 46% of the total manufacturing, and provides employment to 38% of the manufacturing labour force. The biggest challenge Pakistani textile industry is facing today is how to cut down its environmental burdens to cope with the international standards on the issue.

Water heating system as required for dyeing process is one of the major energy consuming areas in fossil fuel-run Pakistani textile industry. Solar water heating, therefore, has a significant contribution towards total environmental impacts associated with textile sector. This case study presents an alternate solution for water heating by means of solar energy. It has been shown herein that under Pakistan's climate solar energy can contribute significantly towards this duty.

At the outset of a collaborative project that has been described by Muneer et al. [57] wherein built-in-storage water heater was identified as a good candidate owing to its low cost and simple construction. To further boost the thermal collection efficiency a new approach was adopted introducing fins within the heater tank (see Fig. 2.37). Fins not only improve the thermal efficiency of the heater but can also add to the stability of heater—fins joined to both the top and bottom surfaces of the heater provide additional strength against hydrostatic forces and can help overcome bulging problems.

Plain and finned water heaters that were expressly constructed were essentially of the same size and specifications except that the latter was provided with fins. Each of these heaters, being of the built-in-storage type, primarily consisted of a water storage tank. For this purpose, two storage tanks were constructed of the size, 1 x 1 m, with a depth of 0.08 m. These tanks were constructed through bending and welding of 1 mm thick stainless steel sheets, which formed the top, bottom, and sides. The top surface of the steel boxes was painted with blackboard paint to act as the absorber plate. Water inlet and outlet valves, each of the size of 0.025 m, were provided on the sides of the boxes at a distance of 0.04 m from the bottom and top respectively. The storage tank needs to be insulated to protect from convection heat loss. For this purpose, the storage tanks were wrapped with 0.04 m of glass wool insulation on all sides and bottom and housed in an outer wooden skeleton box. The top black painted surface of the box was insulated with a 6 mm thick glass sheet that was fixed maintaining a distance of 0.025 m. The steel assembly was used to mount the heaters at an angle of 22°.

The fins used were made of 1 mm thick stainless steel sheet as used in the construction of the heater itself. There were four fins used, each of a length of 0.8 m and of a height of 0.08 m (equal to the depth of the box). These fins were fixed inside the boxes joining the top and bottom surfaces and parallel to its sides, leaving a distance of 0.1 m from both ends as shown in Fig. 2.37. The four fins were fixed with a pitch of 0.2 m. Fins were arc welded at the top surface of the box. However, due to technical problems encountered in the process, it was not a continuous welding contact between the two surfaces but 0.025 m contact of a continuous weld followed by a 0.1 m gap. This provided only a welding contact of 0.16 m, just 20% of the total fin length.

Figure 2.38 shows the temperature distribution of the fin for the present construction, i.e. fin thickness = 1 mm.

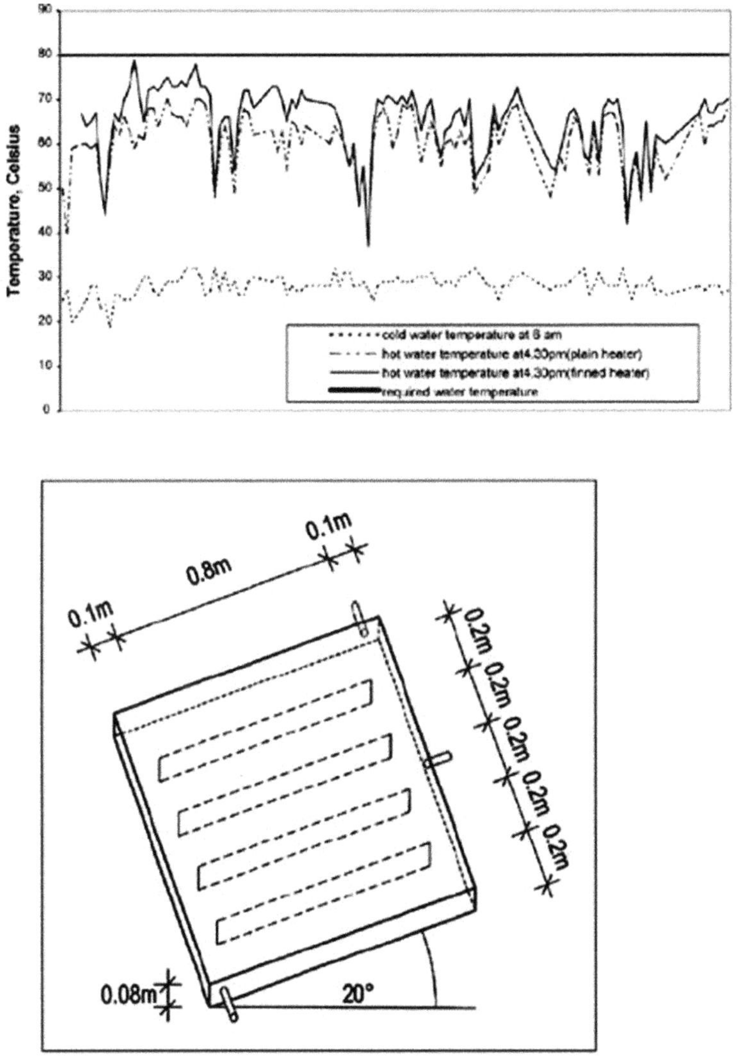

Fig. 2.37 Performance of finned and ordinary built-in-storage water heater (top) and finned-heater design

Comparison of the performance of the finned and plain heaters is included in Fig. 2.37. The finned heater provides higher end-of-day temperature, there being a 10% efficiency increase. A picture of the manufactured heaters is shown in Fig. 2.39.

Figures 2.40 and 2.41 respectively present hourly water temperature profiles for typical clear and overcast days. Note that even on overcast days the energy collection is not seriously hampered.

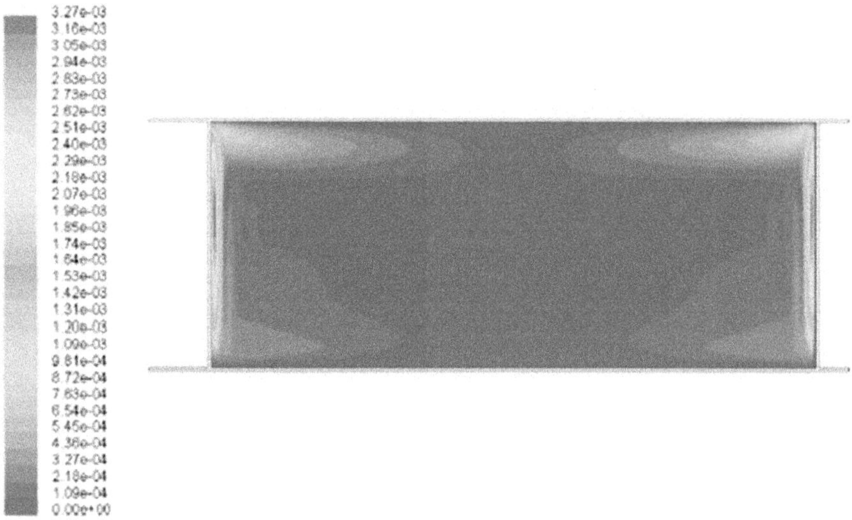

Fig. 2.38 CFD analysis of a finned water heater

Fig. 2.39 Picture of the finned water heater

2.6.7 Solar PV Charging of an Electric Vehicle

This case study presents the experimental findings of a rooftop 2.88 kWp solar PV system that was installed on a private residence in Edinburgh. The electrical consumption of the residence is presented along with the additional electrical load of driving an Electric Vehicle (EV), a Renault Zoe which has a 22 kWh Lithium-ion battery [58].

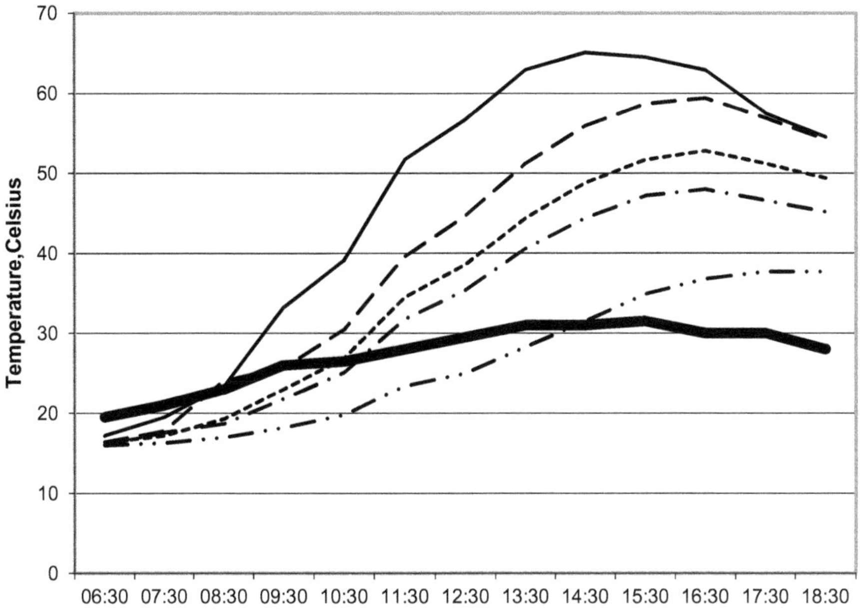

Fig. 2.40 Temperature distribution in finned water heater on a clear day (May 2, 2004). Temperature of the water body at depths of: 1: 260 mm (top curve), 2: 420 mm, 3: 580 mm, 4: 740 mm, 5: 900 mm (bottom curve). Thick, solid line depicts ambient air temperature

The findings may be summarised thus:
Taking into account the embodied energy of the solar PV system, the EV emissions are 21gCO_2e/mile.

The running cost of the EV works out at £88/year, resulting in a net saving of £821/year compared to a petrol-engine car.

A comparison between measured energy generated by the solar PV system and modelling based on the work of Jeffrey et al. [53] was carried out.

A further comparison between experimentally measured and computed energy demand of the EV was made. The purpose-built, VEDEC software enables computation of the EV's traction and regenerative energy. The latter development is described in the works of Muneer et al. [59] and Milligan et al. [60]. The results are presented in Figs. 2.42, 2.43, 2.44, 2.45, 2.46, 2.47, 2.48 and 2.49.

2.6.8 Núñez de Balboa Photovoltaic (PV) Plant

The Iberdrola group in Spain has built the Núñez de Balboa, 500 MWp capacity photovoltaic plant in Badajoz [61].

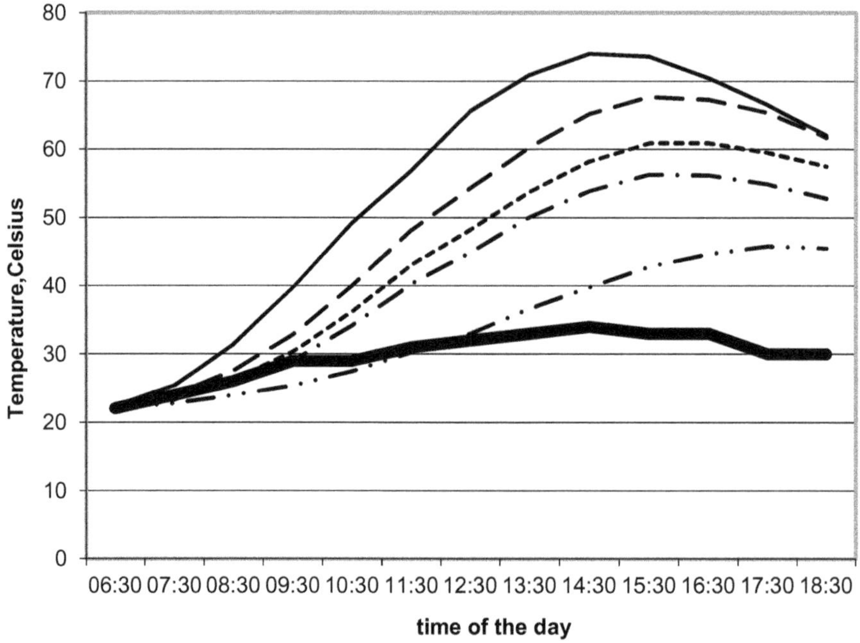

Fig. 2.41 Temperature distribution in finned water heater on an overcast day (May 1, 2004). Temperature of the water body at depths of: 1: 260 mm (top curve), 2: 420 mm, 3: 580 mm, 4: 740 mm, 5: 900 mm (bottom curve). Thick, solid line depicts ambient air temperature

Fig. 2.42 The Newhaven residence in Edinburgh, Scotland with a 2.88 kWp mono-crystalline solar PV array

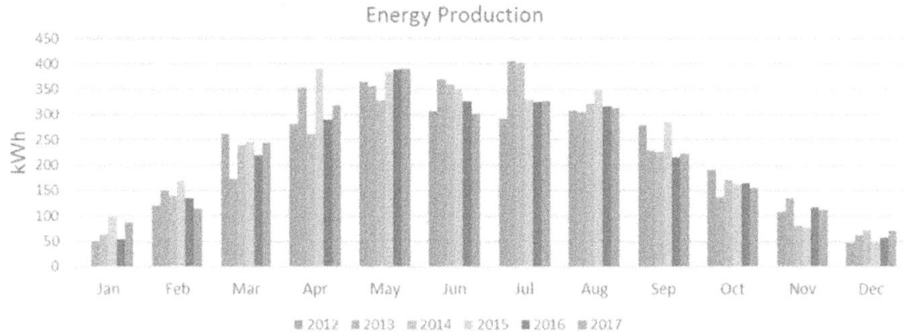

Fig. 2.43 Energy generation profile for the 'Newhaven' house: February 2012–December 2017

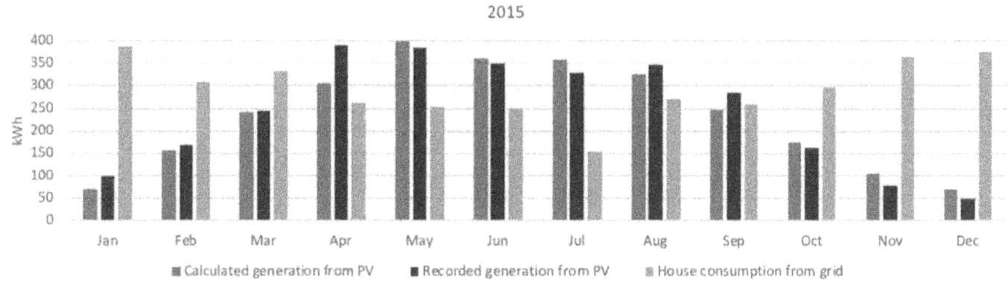

Fig. 2.44 Measured and computed generation and imported energy from the grid for 'Newhaven' house: year 2015

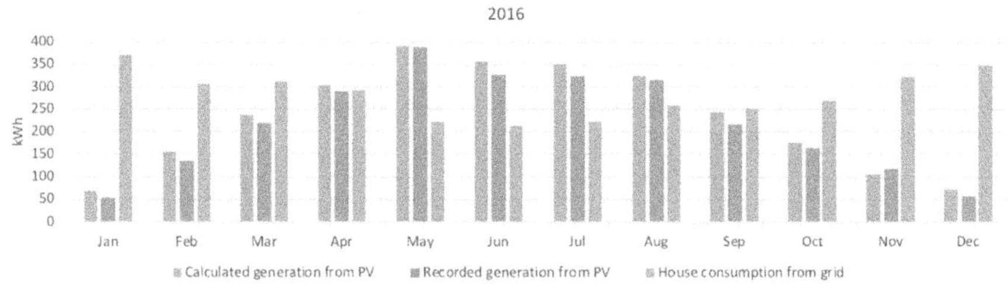

Fig. 2.45 Measured and computed generation and imported energy from the grid for 'Newhaven' house: year 2016

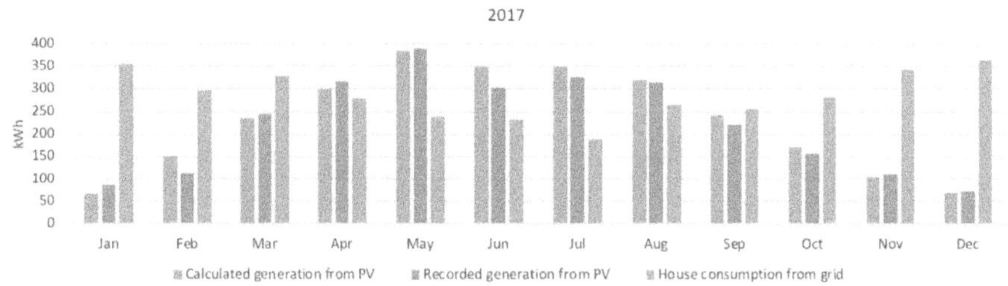

Fig. 2.46 Measured and computed generation and imported energy from the grid for 'Newhaven' house: year 2017

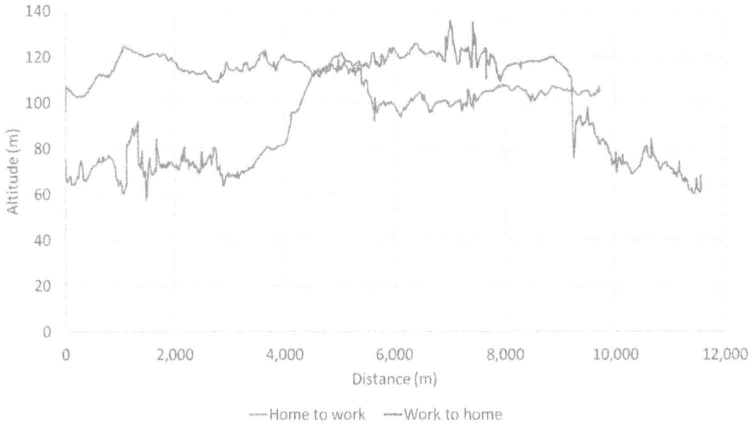

Fig. 2.47 Altitude profile of the route undertaken by the car. This profile is needed to obtain the traction energy required for hill climbing related work

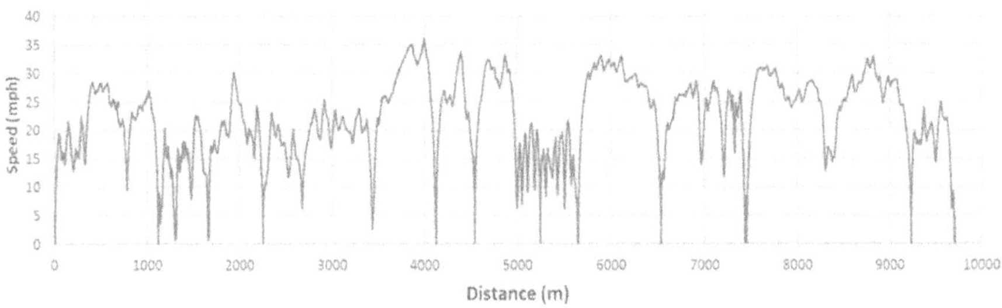

Fig. 2.48 Speed profile of the car. This profile is needed to obtain the energy required for acceleration, wind drag and road friction

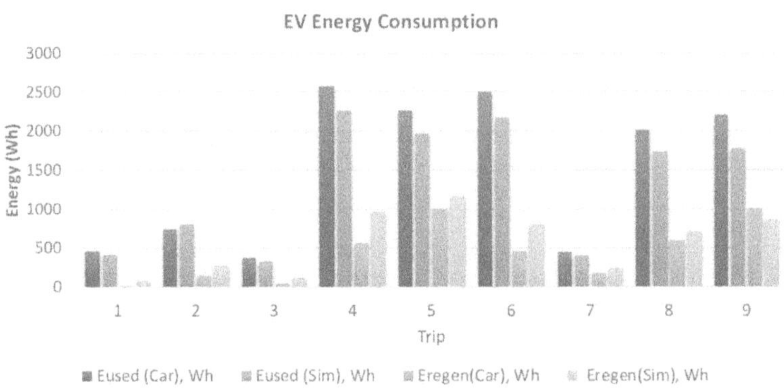

Fig. 2.49 Comparison of energy used for driving the car and energy regenerated—measured and simulated quantities

The plant occupies an area of 1000 ha and has 1.43 million photovoltaic modules. Its electricity generation is 832 GWh/year.

The plant uses 115 inverters and more than 2000 kms of medium and low voltage cables. The plant construction includes a 400 kV electrical substation.

The project was financed with the help of European Investment Bank (EIB) and the Official Credit Institute (ICO) which jointly contributed 290 million euros to finance the plant.

The project is part of Iberdrola's plan to start up 2000 MW of new photovoltaic and wind power plants in Extremadura by 2022—of which more than 1300 MW capacity plants are currently processing. Among the new projects in the autonomous community, some are already in an advanced stage of development, such as the Ceclavín, Arenales, and Campo Arañuelo I and II photovoltaic plants in Cáceres. In addition, the Francisco Pizarro project—which, with 590 MW, will be the largest in Europe. The latter plant will be built in addition to the Majada Alta and San Antonio projects (50 MW each).

The company has plans to relaunch clean energies in Spain, where it plans to invest 8 billion euros by the year 2022.

2.6.9 Plataforma Solar Extremadura Thermal Power Plant

The Plataforma Solar Extremadura is currently the largest solar thermal plant in Spain and one of the largest in Europe. This plant is capable of generating 200 MW electrical power. It is located in Caceres (Extremadura), in the municipality of Logrosán [62].

This plant was built in two phases and consists of four concentrated solar power plants (CSP). Each phase has its own name: Solaben 1, 2, 3, and 6. Each phase has an installed capacity of 50 MW. Solaben 2 and 3 began operating in July and December 2012, respectively, and Solaben 1 and 6 began operating in September 2013.

The two solar thermal plants in the first phase are jointly owned by Abengoa Solar (70%) and Itochu (30%), while the other two are exclusively owned by Abengoa Solar. The cost of construction of the first phase exceeded 500 million euros, with 340 million financed mainly through a loan from SMBC, HSBC, Mizuho, BTMU companies, and the Japanese export credit agency NEXI, while the second phase of the project was carried out with a total investment of 200 million euros.

The company Atlántica Yield is the owner and is in charge of the operation and maintenance of the four plants. The Extremadura Solar Platform is made up of four parabolic trough plants. They occupy a solar field of 440 ha in total. This renewable energy generation complex produces enough energy to supply to more than 100,000 homes.

Each of the four phases consists of 360 parabolic trough collectors measuring 150 m in length, and these include 12 modules. Other infrastructure installations in the largest solar thermal complex in Spain are: a heat transfer fluid (HTF) boiler, thermal oil pipes, a turbine, a steam generator, and cooling towers.

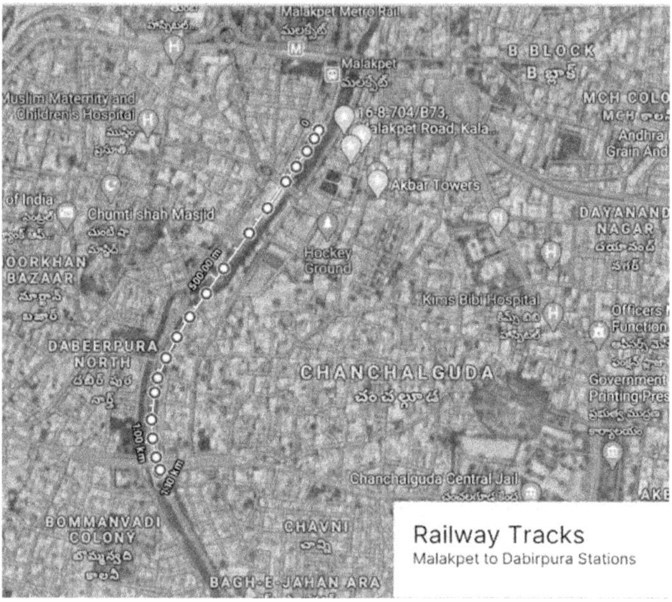

Fig. 2.50 Map of the railway tracks from Malakpet to Dabirpura. The potential for generation of solar PV electricity is provided in Table 2.25. *Credit* Zafar Akber of Xische Holdings of Dubai who further enhance the Google map with distances calculated

2.6.10 Electricity Generation for 'Malakpet Solar Community' Project, Hyderabad, India

The city of Hyderabad in India is a very large metropolis that has been successful in getting inward investment in the form of the creation of a large IT sector. In Greater Hyderabad, there are over a million people employed in the latter sector. The demand for electricity is very high with blackouts and brownouts being quite frequent.

Among the old areas of the city of Hyderabad is the neighbourhood of Malakpet (Latitude: 17.3716° North, Longitude: 78.491° East) which was an estate bestowed upon Malik Yakut, a loyal official who served the seventh Golconda king, Abdullah Qutb Shah of the Qutb Shahi Dynasty of Deccan, that king's reign being from 1626–1672.

Malakpet is located about 4 km south-east of Hyderabad's city centre. It is one kilometre south of 'Musi' river and two kilometres directly south of Kachiguda railway station.

In the early Asaf Jahi times, Malakpet was a village where Iranian relatives of Mir Alam Bahadur (the Prime Minister to the second Nizam of Hyderabad) came and settled. Khairunnisa Begum, the lady in the life of James Kirkpatrick—the then British Resident, belonged to this area. That story has been covered by the Scottish writer, William Dalrymple in his tome 'The White Mughal'.

Table 2.24 Electricity use load profile for an upper middle-class South Indian residence [33]

	Load	Wattage	Quantity	Usage, hours	Energy used, kWh	Day	Night
1	Ceiling fans	75	4	14	4.2	2.1	2.10
2	Exhaust fan	150	1	6	0.9	0.9	
3	Fluorescent lamps	43	7	8	2.408		2.41
4	Incandescent lamp	60	1	5	0.3		0.30
5	CFL lamps	15	6	5	0.45		0.45
6	Television	80	1	6	0.48		0.48
7	Food processor	800	1	1	0.8	0.8	
8	Grinder	750	1	1	0.75	0.75	
9	Washing machine	700	1	1	0.7	0.7	
10	Desktop PC	250	1	8	2	1	1
11	Laptop	45	2	1	0.09	0.09	
12	battery Charger	6	5	6	0.18	0.18	
13	Air-conditioner	1550	2	3	9.3	9.3	
14	Ironing	1100	1	1	1.1	1.1	
15	Mosquito repeller	9	3	12	0.324		0.32
16	Refrigerator	200	1	24	4.8	2.4	2.40
17	Water pump	746	1	1	0.746	0.746	
				Total/day, kWh	**29.5**	**20.1**	**9.50**
				Total/month, kWh	**901**	**612**	**289**

During the seventh Nizam, Osman Ali Khan's time, a special area called 'Araish-e-Baldiya' (City improvement board) was developed on both sides of the Malakpet railway station, which comprised a few hundred houses that were built in the early 1930s.

Table 2.25 Global solar radiation profile for Malakpet, Hyderabad 17.3716° N, 78.491° E, kWh/m^2-day and PV electricity generated, MWh/month

	kWh/m^2-day	MWh/month
Jan	4.979	52.2
Feb	5.857	50.6
Mar	6.079	52.3
Apr	6.664	53.3
May	6.539	50.7
Jun	4.543	34.6
Jul	4.440	35.8
Aug	4.453	37.6
Sep	4.388	38.2
Oct	5.408	50.1
Nov	5.056	50.1
Dec	4.698	50.6

The residents of Malakpet have come up with a plan to generate electricity from solar energy. They have formed a Solar Energy Cooperative with the view of promoting energy sustainability and overcoming the problems of blackouts and brownouts. One of the solutions being pursued is to create a solar corridor by means of installing solar modules along the railway tracks, running from the banks of Musi river, right up to Malakpet- then going beyond to Dabirpura train station. The corridor offers shade-free space which could potentially be used for solar PV electricity generation. Figure 2.50 shows a Google map of the location and the route of the railway which would host PV modules along the tracks.

Table 2.24 and 2.25 respectively show the load profile for a single residence and the monthly-averaged, daily energy that may potentially be delivered using the design configuration shown below. The day-time monthly demand is 612 kWh. Hence the track-side PV installation would be able to support the energy needs of 87 residences in the best month (April).

For the first 885 m run of the railway tracks starting from Malakpet train station end, the PV modules will have an aspect of 126° from North, the subsequent 213 m section of the track will have an aspect of 90°. A respective total of 740 and 180 modules will cover the first and second legs.

References

1. https://www.statista.com/statistics/267358/world-installed-power-capacity/
2. https://www.renewableenergyworld.com/2020/08/24/could-renewables-become-the-top-uk-power-source-of-2020/?utm_medium=email&utm_campaign=2020-08-26&utm_source=rew_weekly_news_letter
3. https://www.vgb.org/en/data_powergeneration.html?dfid=98054

4. DUKES (2019) Digest of UK Energy Statistics, (UK) Department of Business, Energy and Industrial Strategy
5. https://www.iea.org/reports/global-energy-co2-status-report-2019/emissions
6. https://www.evwind.es/2020/07/05/wind-energy-expanded-19-in-2019-with-around-60-gw-of-new-capacity/75563. Accessed on 27 Aug 2020
7. Power-technology.com, https://energy.economictimes.indiatimes.com/news/renewable/worlds-top-10-countries-in-wind-energy-capacity/68465090
8. https://www.renewablepress.com/companies/214-vgb-powertech-e-v
9. https://www.greentechmedia.com/articles/read/siemens-gamesa-takes-worlds-largest-turbine-title
10. https://energynumbers.info/uk-offshore-wind-capacity-factors
11. https://climatenewsnetwork.net/renewables-the-99-9-solution/
12. Celik A, Makkawi A, Muneer T (2010) Critical evaluation of wind speed frequency distribution functions. J Renew Sustain Energy 2:01310
13. https://www.popularmechanics.com/science/a30266828/worlds-cheapest-solar-farm/
14. https://www.globalenergyworld.com/news/sustainable-energy/2020/10/21/sun-cable-earmarks-site-10gw-solar-farm-at-cattle-station-south-darwin
15. https://www.iea.org/reports/world-energy-outlook-2019/electricity
16. Asif M, Muneer T (2007) Energy supply, its demand and security issues for developed and emerging economies. Renew Sustain Energy Rev 11:1388–1413
17. https://www.solarpowerworldonline.com/2019/11/what-are-heterojunction-technology-hjt-solar-panels/
18. Mohanty P, Muneer T, Kolhe M (eds) (2016) Solar photovoltaic system applications a guidebook for off-grid electrification. Springer International Publishing Switzerland
19. Gul M, Kotak Y, Muneer T, Ivanova S (2018) Enhancement of albedo for solar energy gain with particular emphasis on overcast skies. Energies 11:2881. https://doi.org/10.3390/en11112881
20. Muneer T, Ivanova S, Kotak Y, Gul M (2015) Finite-element view-factor computations for radiant energy exchanges. J Renew Sustain Energy 7:033108. https://doi.org/10.1063/1.4921387
21. Muneer T (2004) solar radiation and daylight models. Elsevier, Oxford
22. Langniss O, Ince D (2004) Solar water heating: a viable industry in developing countries. Refocus 5(3):18–21, May-June 2004
23. Renewable Energy (2010) Key facts and figures for decision makers. Global status report. /http://www.ren21.net/gsrS
24. Islam MR, Sumathy K, Khan SU (2013) Solar water heating systems and their market trends. Renew Sustain Energy Rev 17:1–25
25. https://www.gov.uk/government/uploads/system/uploads/attachment_data/file/48188/3147-measure-domestic-hot-water-consump.pdf
26. Renewable Energy Technologies, World Energy Assessment (WEA) Report (2001). www.undp.org/seed/eap/activities/wea/drafts-frame.htmlS
27. Muneer T (1985) Effect of design parameters on performance of built-in storage solar water heater. Energy Conserv Manage 25(3):277
28. Kreider F, Kreith K (1981) Solar energy handbook, McGraw-Hill, New York, Chapter 11:" Solar Water Heating (by G. Skartvedt)". 11.1–11.30, 1981
29. Courtney RG (1976) An appraisal of solar water heating in the UK. BRE Curr Pap 7(76):1976
30. Timilsina GR, Kurdgelashvili L, Narbel PA (2012) Solar energy: markets, economics and policies. Renew Sustain Energy Rev 16(2012):449–465

31. Muneer T (2001) Solar energy: Kempe's engineers year-book. United Business Media, Tonbridge, Kent
32. Jenkins D, Muneer T, Kubie J (2005) A design tool for predicting the performances of light pipes. Energy Buildings 37(5):485–492
33. Zhang X, Muneer T, Kubie J (2002) A design guide for performance assessment of solar light-pipes. Light Res Technol 34(2):149–168
34. Commission Internationale de l'Eclairage (2006) Tubular Daylight Guidance Systems, CIE 173:2006. CIE, Paris
35. powerengineeringint.com
36. https://www.nsenergybusiness.com/projects/talcher-kolar-hvdc-link/
37. https://www.thehindubusinessline.com/economy/north-eastagra-transmission-line-to-be-inaugurated-soon/article9804773.ece
38. Scarlett M, Mortimer J (2001) Fuel cells. In: Kempe's engineers year-book. United Business Media, Tonbridge, Kent
39. Kruger P (2000) Electrical power requirement in the United States for large scale production of hydrogen fuel. Int J Hydrogen Energy 25
40. IHVE Guide C4Flow of fluids in pipes and ducts (1977) The Chartered Institution of Building Services, London
41. Norvelle F (1995) Fluid power technology. West Publishing Company, Minneapolis
42. Odgen J, Steinburger M, Kreutz T (1997) Hydrogen as a fuel for fuel cell vehicles: a technical and economical comparison. In: Eighth annual conference. National Hydrogen Association, VA, 11–13 Mar 1997
43. https://www.globalenergyworld.com/news/sustainable-energy/2020/08/26/north-carolinas-largest-battery-system-now-operating-at-duke-energy-substation?hash=26fc3ab80f98178e725fbd5511876e92&campaignid=73247&messageid=77635&l=2&utm_source=gew&utm_medium=email&utm_campaign=day-z1-n2-hlb&utm_content=btn
44. http://scoreit.online/2020/08/21/tesla-gives-up-crown-for-worlds-biggest-battery/
45. Muneer T, Kolhe M, Doyle A (2017) Electric vehicles: prospects and challenges. Elsevier, Oxford
46. https://www.iea.org/reports/global-ev-outlook-2020
47. https://insideevs.com/news/397240/hydrogen-fuel-cell-sales-2019-7500-globally/
48. https://insideevs.com/news/396177/global-ev-sales-december-2019/
49. www.nextgreencar.com
50. Kato K, Murato A, Sakuta K (1997) An evaluation on the life cycle of photovoltaic energy systems considering production energy of off-grade silicon. Sol Energy Mater Sol Cells 47:95–100
51. Kannan R, Leong K, Osman R, Ho H, Tso C (2006) Life cycle assessment study of solar PV systems: an example of a 2.7 kWp distributed PV system in Singapore. Solar Energy
52. Sweet T, Sweet C, Wu J, Drysdale B, Jenkins N (2013) Case study versus simulation data for a grid-connected 2.45 kWp PV system. In: Presented at the 9th photovoltaic science application and technology (PVSAT-9) conference. Swansea, Wales, 10–12 Apr 2013
53. Jeffrey M, Kelly I, Muneer T, Smith I (2015) Evaluation of solar modelling techniques through experiment on a 627 kWp photovoltaic solar power plant at Edinburgh College—Midlothian Campus, Scotland. J Renew Sustain Energy 7:033128. https://doi.org/10.1063/1.4922450
54. Minutes of the perth and kinross council development management committee, 18 Feb 2015
55. Dowell R (2019) Thermal management for energy efficient agriculture project report. BEng (Honours) project report. Edinburgh Napier University, Apr 2019
56. Department for Business, Energy & Industry, UK, 2018 report
57. Muneer T, Maubleu S, Asif M (2006) Prospects of solar water heating for textile industry in Pakistan. Renew Sustain Energy Rev 10:1–23. https://doi.org/10.1016/j.rser.2004.07.003
58. Heard MA (2018) The supply and demand of energy required for electric vehicles in the UK. BEng (Honours) project report, Edinburgh Napier University, Apr 2018
59. Muneer T, Milligan R, Smith I, Doyle A, Pozuelo M, Knez M (2015) Energetic, environmental and economic performance of electric vehicles: experimental evaluation. Tr Res Pt D 35:40–61. https://doi.org/10.1016/j.trd.2014.11.015

60. Milligan R, Etxebarria S, Muneer T, Gago EJ (2019) Driven performance of electric vehicles in Edinburgh and its environs. Energies 12:3074. https://doi.org/10.3390/en12163074, 2019
61. https://www.iberdrola.com/conocenos/lineas-negocio/proyectos-emblematicos/planta-fotovoltaica-nunez-de-balboa
62. http://www.abengoa.es/web/es/compania/index.html

Chapter 3
Wind Energy and Solar PV Developments in China

3.1 Introduction

Energy demand is expected to increase significantly in the coming years due to population growth and economic development. The largest increases in energy demand will be in developing countries, where the global share of energy consumption is forecast to rise from 46 to 58% between 2004 and 2030. Along with its rapid economic growth over the past three decades, China has become the largest carbon emitter since 2007 and the largest energy consumer since 2009 [1].

In 2020, China's total energy consumption reached 7510 TWh, increasing by 3.1% over the previous year and 5.7% over the average annual growth rate during the 13th Five-Year Plan period [2].

The current energy situation, which is dominated by fossil fuels, has led to a number of problems. The first problem created in China is the growing and severe air pollution [3]. In 2007, the year in which the concept of "ecological civilisation" was first introduced at the political level in China, only 1% of Chinese urban residents breathed air that the European Union considered safe [4]. In recent years, an increasing number of hazardous pollution incidents, represented by the haze and haze that have affected most of the country's eastern and northern regions since early 2012, has not only posed a serious threat to citizens' health but have also hindered China's sustainable growth [5]. As Poon et al. (2006) have pointed out, the combustion of fossil fuels, especially coal, is a major cause of mortality. In recent years, the increase in air pollution has also attracted international concern and thus damaged China's international image [6].

The second major problem related to energy consumption is the emission of greenhouse gases (GHG) generated by the combustion of coal. Humans are currently emitting CO_2 into the atmosphere at a rate a hundred times greater than at any time in human history before the onset of industrialisation. More than half of the CO_2 emitted into the atmosphere has been emitted in just the last 30 years. The United States was the biggest polluter until 2011, a leadership that was then taken over by China due to its industrialisation process, which began with the policy of reform and

T. Muneer et al., *The Coming of Age of Solar and Wind Power*, Green Energy and Technology, https://doi.org/10.1007/978-3-030-92010-4_3

opening up, but with effects on pollution since the 1990s. The global greenhouse gas pollution problem currently comes from China (26.6%), the United States (13.1%), India (7.1%), and Russia (4.6%) [7].

The third problem is that China is increasingly dependent on imported coal and energy. Although it is the world's largest coal producer, it has been a net importer of coal since 2007 and of oil and gas since 1990. In 2013, China's coal import dependency ratio reached an all-time high of 8.1%, with net coal imports amounting to approximately 320 million tonnes. Overall, the ratio of net imported energy to total energy consumption has increased rapidly from 3% in 1998 to 11% in 2011 [8].

Fortunately, China has recognised these problems and has started to make efforts to control this situation. During the 2014 Asia Pacific Economic Cooperation (APEC) summit, China and the United States issued a joint announcement on climate change. According to the announcement, China pledged to reach its peak CO_2 emissions and increase the share of non-fossil fuels in primary energy consumption to approximately 20 per cent by 2030.

Renewable energy sources, being clean and environmentally friendly, have a clear advantage over other energy systems. Currently, the renewable energy sector covers 13.5% of the world's energy demand. Since the beginning of this century, China has been trying to meet the challenge of developing renewable energy, particularly wind and solar energy, and remarkable achievements have been made in these two sectors. The National Energy Administration (NEA) reported 68.6 GW of onshore wind installations connected to the grid in 2020, bringing the total onshore installations to more than 272 GW [9].

Compared to its wind power market, China's domestic solar PV market has been smaller. However, its solar PV manufacturing industry has grown dramatically since 2004, now ranking first in the world in terms of production scale, which is due to the incentive policies of the Chinese government [10].

This chapter presents China's current energy landscape in terms of its total energy demand and supply. Solar, wind, and thermal energy have been compared in terms of its current economics.

3.2 Geographical Background

China is located in East Asia and has an area of 9,600,013 km^2. The People's Republic of China is the third largest country in the world (Fig. 3.1) [10].

China has great physical diversity. The country's eastern plains and southern coasts are made up of fertile lowlands and hills. Most of the country's agricultural production and population is concentrated here. The southern parts of the country (south of the Yangtze River) are hilly and mountainous. The west and north of the country are dominated by sunken basins (such as the Gobi and Taklamakan), rolling plateaus, and towering massifs. It contains part of the highest plateau on earth, the Tibetan Plateau [11, 12]. East and South China stretch along the Pacific Ocean, with the South China Sea to the south and the East China Sea and Yellow Sea to the east.

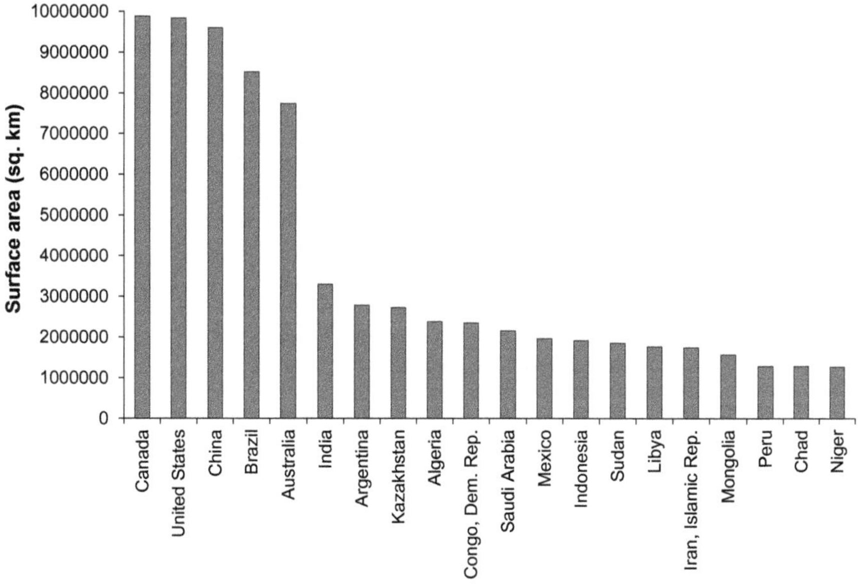

Fig. 3.1 Surface area (sq. km). *Source* Prepared and adapted by the author on the basis of https://data.worldbank.org/indicator/AG.SRF.TOTL.K2

The topography is diverse, with snow-capped mountains, deep river valleys, broad basins, high plateaus, rolling plains, terraced hills, sandy dunes, with many other geographical features and other landforms present in countless variations. In general, the land is high in the west and slopes down to the east coast. Mountains (33%), plateaus (26%), and hills (10%) account for almost 70% of the country's land area. Most of the country's arable land and population is located in the lowland plains (12%) and watersheds (19%), although some of the larger watersheds are filled with deserts [13, 14].

China has been divided by the Chinese government into five homogeneous physical macro-regions [12, 13]:

- Eastern China.
  - Northeast plain.
  - Northern plains.
  - Southern hills.
- Xinjiang-Mongolia.
- Tibetan plateau.

Eleven of the 17 highest peaks on Earth are on China's western borders (Fig. 3.2). These include the world's highest peak, Everest (8848 m), in the Himalayas on the border with Nepal, and the world's second-highest peak, K2 (8611 m), on the border with Pakistan. North of the Himalayas and east of the Karakorum/Pamir is the vast

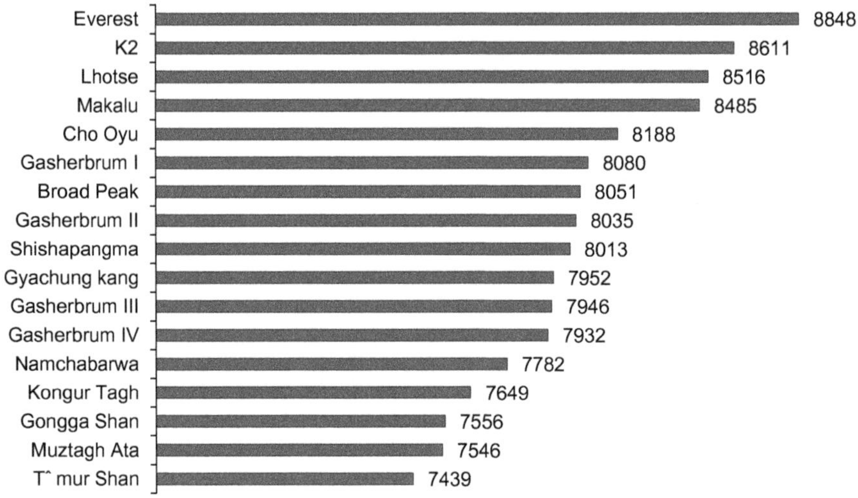

Fig. 3.2 Tallest mountain peaks of China. *Source* Prepared and adapted by the author

Tibetan Plateau, the largest and highest in the world, also known as the "Roof of the World". The plateau has an average elevation of 4000 m above sea level and covers an area of 2.5 million square kilometres or one-fifth of China's landmass [14, 15].

The People's Republic of China is located at the geographical coordinates of 35° 51′ 42″ N and 104° 11.724′ E. The distinctive feature of China's climatic conditions is its variety. The northern region has a "subarctic" climate, while the southern part is dominated by a tropical climate. The latitude and longitude of Beijing, the capital of China, are 39° 56′ N and 116° 24′ E respectively. Monsoon winds, caused by differences in the heat-absorbing capacity of the mainland and the ocean, dominate the climate. During the summer, the East Asian monsoon transports warm, moist air from the south and provides most of the annual rainfall over much of the country [15, 16]. In contrast, the Siberian anticyclone dominates during the winter, providing cold and comparatively dry conditions. The advance and retreat of the monsoons largely determine the timing of the rainy season across the country. Although most of the country lies in the temperate belt, its weather patterns are complex [17–21].

Temperature differences in winter are considerable, but in summer the variation is much smaller [22–24]. Rainfall is almost always concentrated in the warmer months, although annual totals range from less than 20 mm in northwest Qinghai and the Turpan Depression in Xinjiang to easily over 2000 mm in Guangdong, Guangxi, and Hainan [25–28]. Only in parts of the Dzungaria region of Xinjiang is the striking seasonal variation in rainfall that defines the Chinese (and, to a large extent, East Asian) climate absent [19–21, 29].

Annual sunshine duration ranges from less than 1100 h in parts of Sichuan and Chongqing to more than 3400 h in northwest Qinghai. Seasonal sunshine patterns vary considerably by region, but in general, the north and the Tibetan Plateau are sunnier than the south of the country [30].

3.3 Historical Background

In the late nineteenth and early twentieth centuries, rural areas of the country maintained a system of communes, the collective farms that were the mainstay of Maoism [31–33]. By the mid-1980s, Deng Xiaoping's reform and opening-up process had achieved among its advances the free market for agricultural products. Factories had more autonomy. On the coast, the first special economic zones had been launched. Foreign trade was taking off, rising from 10% of GDP in 1980 to 35% in 1986. Nationwide, GDP per capita was $1154. Beijing then had a population of 9.8 million. These opening-up reforms created an economy totally dependent on foreign demand for low value-added goods [34–37].

In this context of political and economic openness, corruption began to skyrocket and, above all, a division between rich and poor began to take shape which, 30 years later, is one of the country's main ills. As a result, the Tiananmen student movement emerged in 1989 [38, 39].

Jiang Zemin, Deng's successor, continued on the path of reform [40]. Its prime minister, Zhu Rongji, laid the foundations for what was to be the spectacular growth of the next decade. Thousands of state-owned enterprises were privatised, inflation was brought under control, and bureaucracy was reduced. Gradually, China's productive fabric was scaled up towards the manufacture of higher value-added goods (chips, telephones, medical devices…) that would generate higher profits and reduce dependence on foreign trade [41].

In 2001, after years of negotiation, China took a new leap in its economic development: the still nominally communist People's Republic joined the World Trade Organisation [42]. By 2005, China had become the world's factory thanks to a huge labour force, low wages, and markets eager to sell resources to China and buy its products.

After the 2008 global crisis, the Chinese authorities put forward a new strategy that included expanding domestic demand (increasing imports and reducing the external deficit) to achieve stronger and more sustainable growth over time. This has led China to consume millions of foreign goods and services, supporting the growth of other countries. This year the country's domestic consumption accounted for 35.3% of China's GDP [43, 44].

China now wants to start producing a greater share of all the high value-added goods and services it now imports, reducing foreign dependence. In this new model, investment is going to new destinations, such as the service sector, technology, and innovation. Premier Li Keqiang has given his public support to Chinese startups [45, 46]. The Made in China 2025 plan includes industrial restructuring to promote high value-added products in sectors such as biomedicine and robotics [47–49]. The country's leaders now want to take the final step to produce such goods, the demand for which was largely met by imports from the countries that produce these goods.

3.4 Socio-economics of China

The energy availability of any country has a strong relationship with its economic and social stability. Per capita energy consumption is a measure of intensity, resulting from the ratio of domestic energy consumption to population. It is an index used to measure the prosperity of any society. China's per capita energy consumption for the year 2018 was 4624.7 kWh (Fig. 3.3), ranking 46th among the world's nations with the highest per capita consumption [50].

China is now the most populous country in the world [11], closed 2020 with a population of 1,439,323,774 people. It has an average population density of 146 inhabitants per km^2. The country is divided into 33 administrative entities: 22 provinces, five autonomous regions, four municipalities, which are under the direct administration of the central government, and two special administrative regions [12].

The 22 provinces are as follows: Anhui, Fujian, Gansu, Guangdong, Guizhou, Hainan, Hebei, Heilongjiang, Henan, Hubei, Hunan, Jiangsu, Jiangxi, Jilin, Liaoning, Qinghai, Shaanxi, Shandong, Shanxi, Sichuan, Yunnan, and Zhejiang.

The five autonomous regions are: Zhuang of Guangxi, Inner Mongolia, Hui of Ningxia, Uyghur of Xinjiang, and Tibet, the four municipalities are: Beijing, Chongqing, Shanghai, and Tianjin. Special administrative regions are Hong Kong and Macao.

The five autonomous regions are associated with the five major minorities: Tibetans, Uighurs, Mongols, Hui, and Zhuang. The metropolitan areas of Beijing, Tianjin, Shanghai, and Chongqing are the municipalities under direct administration

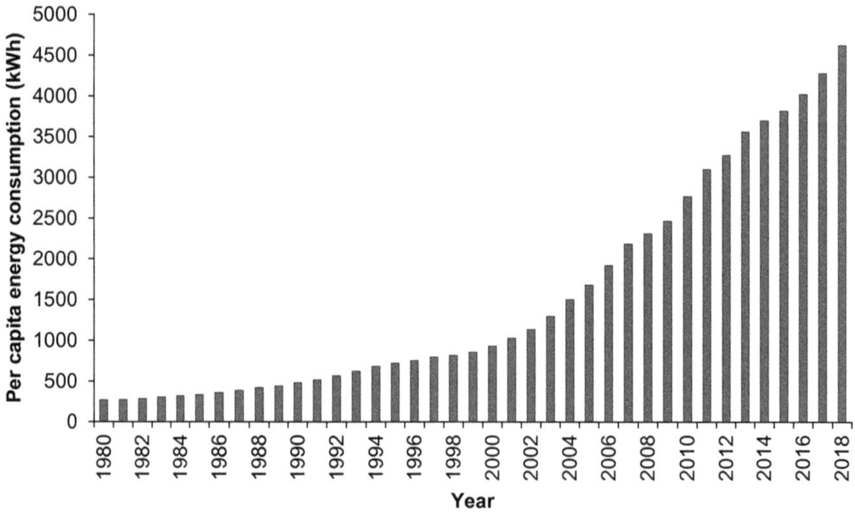

Fig. 3.3 China's per capita energy consumption (kWh). *Source* Prepared and adapted by the author from https://datosmacro.expansion.com/

of the central government; and the former European colonies of Hong Kong and Macao maintain a large measure of autonomy as special administrative regions [12].

China is the world's second largest economy in terms of Gross Domestic Product (GDP) [51, 52]. As can be seen in Fig. 3.4, GDP has experienced sustained and rapid growth. For the year 2019, GDP per capita was 9164 euros. Its public debt in 2019 was €6,769,837 million, with a debt of 52.63% of GDP. Its per capita debt is 4,844 euros per inhabitant [53–55]. The latest annual rate of change of the Consumer Price Index (CPI) published in China is for February 2021 and was −0.2%, 1 tenth of a percentage point higher than in the previous month. The monthly change of the CPI was 0.6% so that the cumulative inflation in 2021 is −0.3% [53].

China has a highly diversified economy, dominated by the manufacturing and agricultural sectors, being one of the largest producers and consumers of agricultural products. It is estimated that the agricultural sector employed 24.7% of the labour force in 2020 [56] and accounted for 7.1% of GDP, although only 15% of China's land (about 1.2 M km^2) is arable. China is the world's leading producer of cereals, rice, cotton, potatoes, and tea. In terms of livestock, it also dominates sheep and pig farming and world fish production [57–59].

In addition, the country is rich in natural resources and has significant coal reserves, which account for two-thirds of total primary energy consumption. China is the world leader in the production of certain minerals (tin, iron, gold, phosphates, zinc, and titanium) and has significant reserves of oil and natural gas, making the

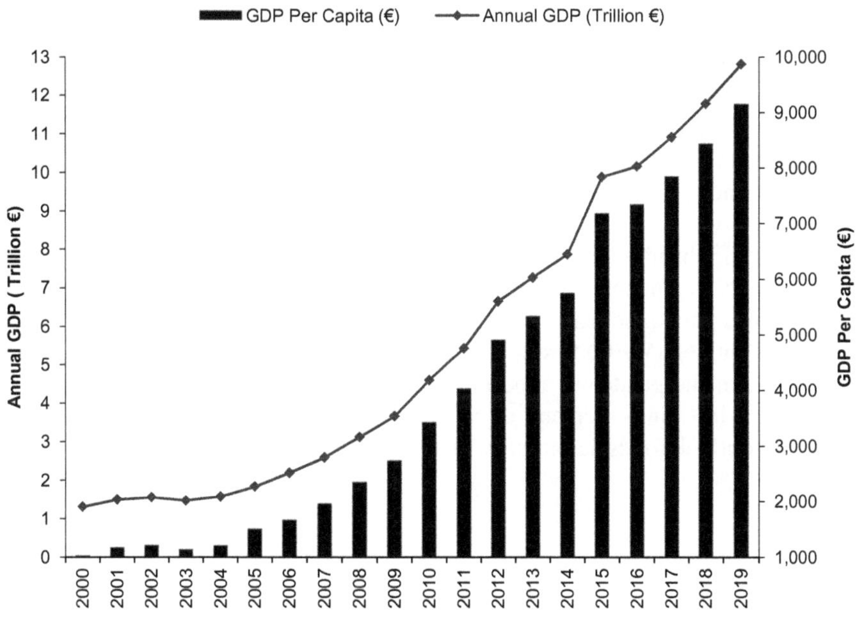

Fig. 3.4 Annual GDP and GDP per capita (2019). *Source* Prepared and adapted by the author from https://data.worldbank.org/indicator/NY.GDP.MKTP.KD.ZG?locations=CN

country the world's fifth-largest oil producer, with 4.89 million barrels per day in 2020 [60].

The industry sector contributes to approximately 39% of China's GDP and employed 28.2% of the population in 2020 [57]. China has become one of the preferred destinations for outsourcing global manufacturing units due to its cheap labour market, despite rising labour costs in recent years. China's economic development has coincided mainly with the development of a competitive and outward-oriented manufacturing sector. More than half of China's exports are made by foreign-owned firms. Their share in the sector's value added varies by industry: more than 60% for electronics and less than 20% for most producer goods. The state-owned sector still contributes about 39% to GDP [60–62].

The share of the service sector in GDP is about 54% and employs about 47% of the labour force [57]. Although the sector's share of GDP has been growing in recent years, the services sector as a whole, encumbered by public monopolies and restrictive regulations, has not progressed. The sector's development has been constrained by the country's focus on manufactured exports and substantial barriers to investment in the sector. However, the Chinese government has been focusing more on the service sectors recently, particularly in sub-sectors such as finance, logistics, education, health, and also aims to be among the top exporters in transport, tourism, and construction [61, 62].

China's importance is even greater in the energy sector. Since 2009, China has become the world's largest energy consumer [63]. If China's GDP represents 16.2% of the world total, and its population 18.3%, its primary energy consumption reaches 23.6%, and its CO_2 emissions 27.2% [57, 64]. In some sub-sectors, in fact, its weight is even greater: China absorbs more than 50% of global coal consumption and is also the largest importer of crude oil, as well as the second largest consumer after the US. In other words, in geopolitical terms, the Chinese energy sector is a decade ahead, in the sense that China has the weight it will have in all other economic sectors by 2030 if current trends continue. Rapid growth is one of the most pronounced features of China's energy.

On the other hand, the COVID-19 pandemic had a strong impact on the global economy in 2020. The International Monetary Fund (IMF) estimates a 4.9% contraction in global economic activity during 2020, 1.9% below the April 2020 World Economic Outlook (WEO) forecast, and worse than the global financial crisis [65]. The impact of the pandemic appears to have affected both sides of most sectors and markets in China—demand disruptions met with supply problems—and the short-term outlook for agriculture, industry, and services is uncertain.

3.5 Energy Resources

China plans to rely on more electricity generation from nuclear, renewable, and natural gas sources to replace some coal-fired generation, in order to reduce carbon dioxide emissions and heavy air pollution in urban areas.

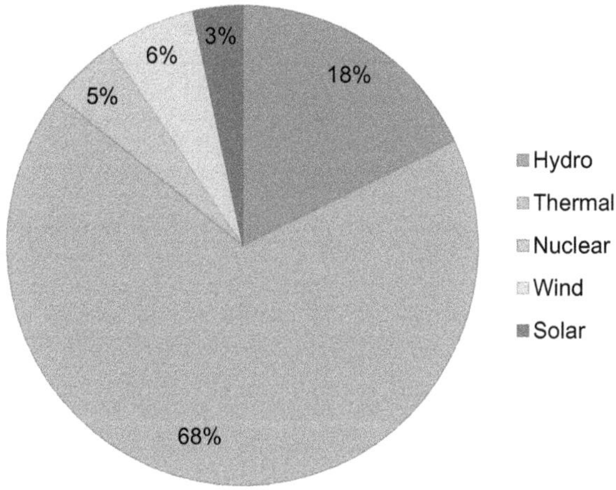

Fig. 3.5 Power generation (TWh) in China (2020). *Source* Prepared and adapted by the author from https://english.cec.org.cn/detail/index.html?3-1128

By the end of 2020, China's installed generation capacity reached 2.2 TW, up 9.5% year-on-year. During the 13th Five-Year Plan period, the average annual growth of installed generation capacity was 7.6%. In particular, China's installed hydro, thermal, nuclear, grid-connected wind, and solar power generation capacity reached 370 GW, 1250 GW, 49.89 GW, 280 GW, and 250 GW, respectively (Fig. 3.5). The installed capacity of non-fossil power amounted to 980 GW, representing 2.8% more than the previous year. Installed coal capacity reached 1080 GW, falling below 50% for the first time [2].

On the other hand, China's power generation reached 7620 TWh, up 4.0% year-on-year. During the 13th Five-Year Plan period, the average annual growth of China's power generation reached 5.8%, in which non-fossil power generation grew at an average annual rate of 10.6%, up 6.7% points. Fossil power generation grew at an average annual rate of 3.5%, from 67.9 to 60.8%. Specifically, hydro, thermal, nuclear, grid-connected wind and solar power generation reached 1360 TWh, 5170 TWh, 366.2 TWh, 466.5 TWh and 261.1 TWh, respectively, up 4.1, 2.5, 5.0, 15.1 and 16.6% year-on-year [2].

According to the China Electricity Council, China's installed generation capacity in 2021 is expected to reach 2.37TW, up 7.7% from 2020. Non-fossil generation capacity will reach 1.12TW, accounting for 47.3% of total installed capacity, 2.5% points higher than the base figure at the end of 2020. Wind and solar power will increase by three percentage points, highlighting the growing need for flexibility and regulatory capacity in the electricity system [2].

3.5.1 Petroleum and Other Liquids

Although China was the world's fifth-largest producer of oil and other liquids in 2019, most of the country's production comes from fields that require costly enhanced oil

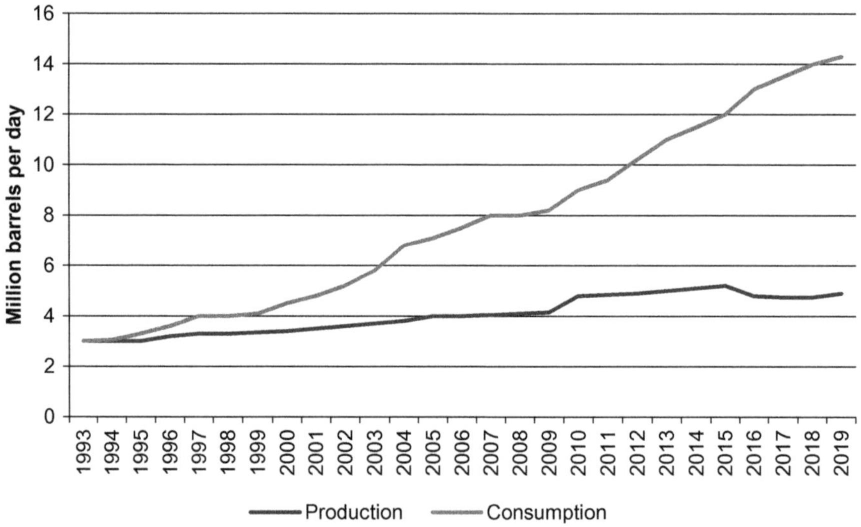

Fig. 3.6 China's petroleum and other liquids production and consumption, 1993–2019. *Source* Prepared and adapted by the author from https://www.eia.gov/outlooks/steo/pdf/steo_full.pdf

recovery techniques to sustain production. After declining sharply for three years, oil and other liquids production in China reversed course and increased to 4.9 million barrels per day (b/d) in 2019 (Fig. 3.6). Nearly 80% of total liquids production came from crude oil, with the remainder from coal and methanol-to-liquids conversions, biofuels, and refinery processing gains.

Oil production from coal-to-liquids (CTL) plants was estimated at 108,000 b/d and methanol-to-liquids at about 500,000 b/d in 2019 [8]. In late 2016, Shenhua Group commissioned the world's largest CTL plant, Ningxia, with capacity to produce more than 80,000 b/d of oil [66].

3.5.2 Natural Gas

China has gigantic natural gas reserves, estimated at 50 trillion cubic metres (tcm), most of which would be unconventional reserves (about 15% of the world's unconventional gas reserves are in China). The Tarim basin (in the western province of Xinjiang) or the Sichuan basin are among the most important gas fields in China [67].

China's natural gas production has steadily increased over the past few years as the country seeks to meet the growing need for natural gas. China's NOCs (National Oil Corporation) produced 0.18 trillion cubic metres of natural gas in 2019, up 8% from 2018 (Fig. 3.7) [8]. Although still in its early stage of development, China's

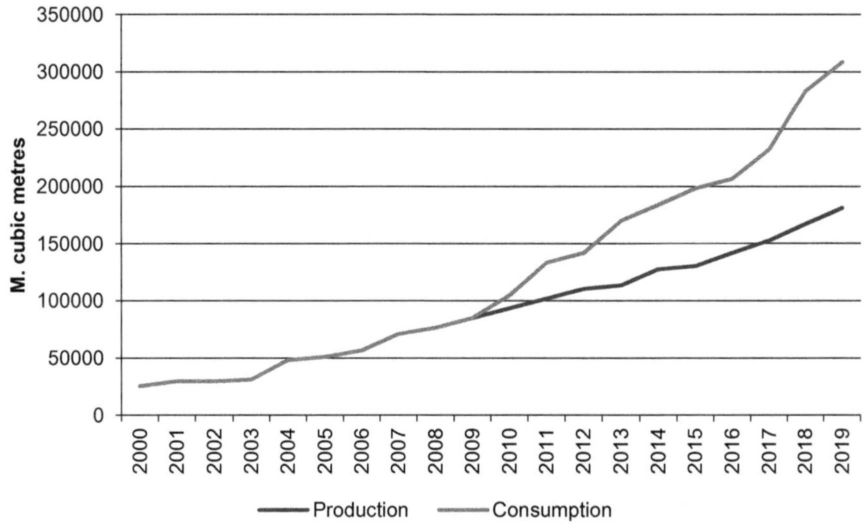

Fig. 3.7 China's dry natural gas production and consumption, 2000–2019. *Source* Prepared and adapted by the author from https://www.eia.gov/outlooks/steo/pdf/steo_full.pdf

shale gas production increased substantially by 14% from 2017 levels to around 10.4 billion cubic metres in 2018 [8].

Offshore natural gas production increased by 10.5% from 2018 to reach 9.5 billion cubic metres in 2019, mainly due to growth in the South China Sea [68]. CNOOC, China's leading offshore producer, plans to bring on stream the country's second deepwater natural gas field, Lingshui 17-2, and the newly explored large Bozhong 19-4 natural gas and condensate field in northeast China's Bohai Bay by 2022 [69].

3.5.3 *Coal*

Coal production, which declined for three consecutive years until 2016, has increased every year since then and reached an estimated 4.1 billion short tons in 2019 (Fig. 3.8) [70]. The Chinese government adopted a supply-side approach to control the volatility of domestic coal prices through a target price range, which allowed domestic producers to be profitable and compete with coal imports.

As coal demand from the power sector and coal prices began to rise in late 2016, China relaxed several policies restricting domestic coal production. China increased production capacity and started to replace uncompetitive and obsolete mining capacity. Production growth came from the largest coal-producing mines in the north-central and northwestern parts of the country. China continues to replace obsolete coal capacity with new and more efficient mining capacity and to close

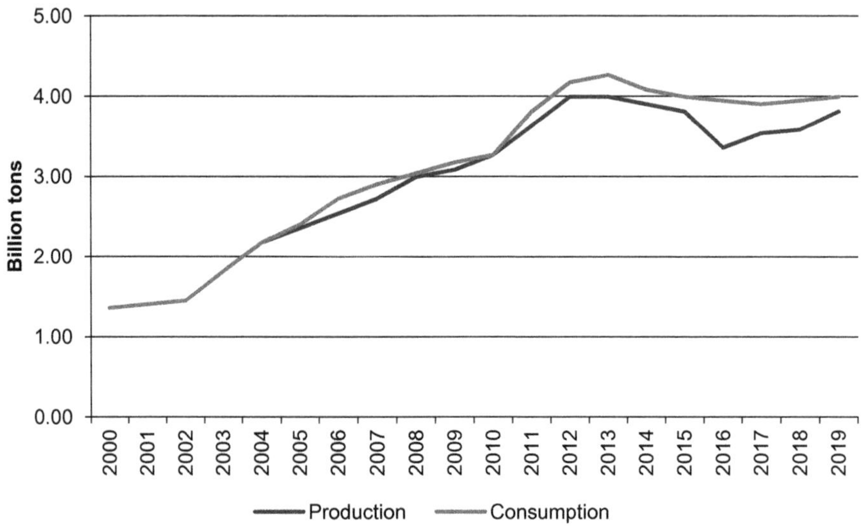

Fig. 3.8 China's coal production and consumption, 2000–2019. *Source* Prepared and adapted by the author from https://www.eia.gov/outlooks/steo/pdf/steo_full.pdf

smaller mines in the eastern and southern regions of the country [71]. In addition, China's recent expansion of long-haul rail capacity, such as the Haoji railway launched in October 2019, to connect inland coal production centres with eastern demand centres, is critical to bolster domestic production and respond to coal demand [72].

3.6 Energy Demand

In 2020, China's total energy consumption reached 7510 TWh, increasing by 3.1% year-on-year and 5.7% over the average annual growth rate during the 13th Five-Year Plan period (Fig. 3.9). The steady economic recovery was the main driver of growth. For primary industry, energy consumption increased by 10.2% to 85.9 TWh. This rapid growth of primary industry consumption is mainly attributed to China's effort to improve rural power grids and alleviate poverty, which boosted the development of rural areas and released the energy consumption potential of primary industry. The energy consumption of secondary industry reached 5120 TWh in 2020, an increase of 2.5% year-on-year. Improved work resumption drove this growth. Tertiary industry consumption reached 1210 TWh, up 1.9% yoy. With the return to normal business and market operations, tertiary industry consumption growth increased quarter-on-quarter. In particular, energy consumption of the information services and software industry grew by 23.9% year-on-year due to the rapid deployment of new technologies such as big data, cloud computing, and the Internet of Things, which

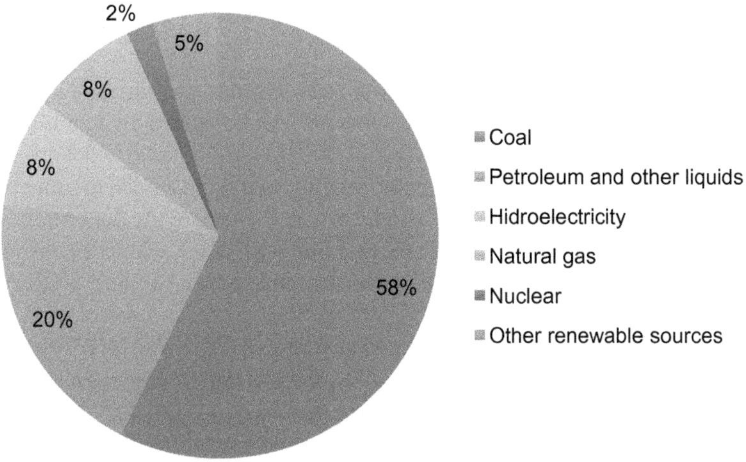

Fig. 3.9 China total primary energy consumption by fuel type (2019). *Source* Prepared and adapted by the author from https://www.bp.com/content/dam/bp/business-sites/en/global/corporate/pdfs/energy-economics/statistical-review/bp-stats-review-2020-full-report.pdf

contributed to the rapid growth of online industries such as job shifting, life service, culture and entertainment, and online education. Residential energy consumption reached 1090 TWh, up 6.9% year-on-year. Falling temperatures in December have contributed to increased demand for heating and electricity, leading to rapid growth in residential electricity consumption in the fourth quarter [73] (Fig. 3.10).

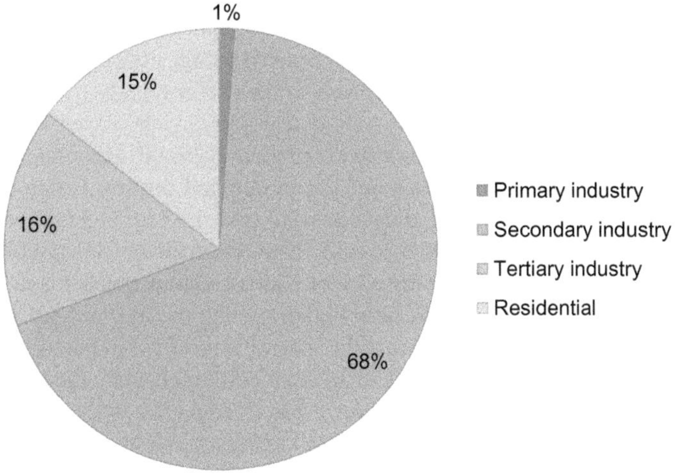

Fig. 3.10 Electricity demand (TWh) in China (2020). *Source* Prepared and adapted by the author from https://www.bp.com/content/dam/bp/business-sites/en/global/corporate/pdfs/energy-economics/statistical-review/bp-stats-review-2020-full-report.pdf

3.6.1 Petroleum and Other Liquids

China's oil consumption growth accounted for an estimated two-thirds of incremental global oil consumption in 2019. China consumed an estimated 14.5 million b/d of oil and other liquids in 2019, an increase of 500,000 b/d, or nearly 4%, over 2018 (Fig. 3.6) [8]. As can be seen in Fig. 3.6, the growth in demand has led to a much larger imbalance between consumption and production in the face of the flat profile of total production, making China the world's largest importer of oil. Chinese consumption amounted, as mentioned above, to 14.5 million barrels per day against production of barely 5 million barrels per day.

Diesel and gasoline accounted for the largest shares (27% and 24%, respectively, in 2018) of oil products consumed over the past few years. However, the pace of oil demand growth in the transport sector has slowed in recent years due to China's economic slowdown, stricter environmental measures resulting in tighter fuel efficiency standards and restrictions on the use of urban vehicles, and increased penetration of alternative fuel vehicles (electric vehicles, compressed natural gas vehicles and trucks and trains running on liquefied natural gas).

3.6.2 Natural Gas

Several factors have contributed to the strong growth in natural gas consumption in recent years. Poor air quality, particularly in China's northeastern urban areas, where excessive use of coal in winter causes smog and dangerous levels of pollution, led the government to impose fuel switching from coal to natural gas for industrial use, power generation, and residential and commercial heating. Strict environmental targets set by the central government in 2017 created a surge in demand for natural gas by northern cities during peak demand usage in the winter months of 2017 and 2018 [2]. Other drivers of demand were low natural gas prices, increased use of natural gas in the transport sector, and the expansion of natural gas infrastructure to alleviate supply bottlenecks and transport natural gas to demand centres. China relaxed its coal-to-gas switching programme in late 2018 to ease the natural gas shortages that occurred in the winter of 2017–18, especially in northern cities. This policy shift and slowing economic growth caused the pace of China's natural gas demand growth in 2019 to decelerate from the significantly higher growth in 2017 and 2018 [74].

China's government plans to increase the share of natural gas as part of total energy consumption from nearly 8% in 2019 to 14% in 2030 to alleviate the high levels of pollution resulting from the country's heavy use of coal [75]. Although natural gas remains a small contributor to China's overall energy portfolio, the country is now one of the fastest growing natural gas markets in the world.

China's natural gas consumption rose 9% in 2019 to 10.8 Tcf, up from 9.9 Tcf in 2018 [8]. Over the last decade, China's demand for natural gas increased rapidly by around 13% per year, making it the world's third largest consumer of natural gas, behind the United States and Russia (Fig. 3.7) [8]. While the majority of natural gas consumption comes from industrial users, including mining and oil and natural gas extraction (which accounted for more than 40% in 2018), the shares of natural gas consumption in the electric power and transportation sectors have increased over the past decade [76]. Natural gas is generally more expensive than coal and continues to face supply constraints due to insufficient storage, import terminals, and pipeline capacity. The pace of demand growth in the coming years is expected to depend on China's environmental policies in the energy sector and increased coal-to-gas switching in households and industries. The pace of demand growth is also likely to depend on the rate at which China can build its natural gas infrastructure. The response to the COVID-19 pandemic is expected to negatively affect natural gas demand growth in China, especially in the industrial, commercial, and transportation sectors [77].

3.6.3 Coal

After several years of declines, China's coal consumption grew by 1% each year in 2018 and 2019, based on physical volume (over 4.3 billion short tons in 2019), according to estimates by China's National Bureau of Statistics (NBS) (Fig. 3.8) [71].

The power sector accounted for nearly 60% of China's coal consumption in 2018, with the remainder of China's coal use coming from industry, such as steel and cement production, and residential heating [78].

China's coal demand in the coming years is likely to be determined by the magnitude of the COVID-19 pandemic in China and its impact on electricity and industrial demand growth, as well as the government's continued policy on air quality and fuel switching. Coal will eventually disappear from the residential heating sector as the government aims to switch more and more households from coal to natural gas or electric boilers each year. Competition from cleaner fuels, China's shift to a less energy-intensive economy, and the trade conflict with the United States will pose downside risks to coal demand in the coming years. However, coal is likely to remain a mainstay of power and heating demand due to the country's abundant resources and the government's intention to increase the use of clean coal technology.

3.7 Renewable Energy Prospect in China

China's economy has experienced strong growth over the last three decades with the consequent degradation of the environment due to excessive carbon dioxide (CO_2) emissions. China alone pollutes as much as the United States, India, Russia, and Japan combined. CO_2 emissions during 2019 have grown by 378,129 kilotonnes, 3.39% compared to 2018. CO_2 emissions in the same year were 11,535,200 kilotonnes,

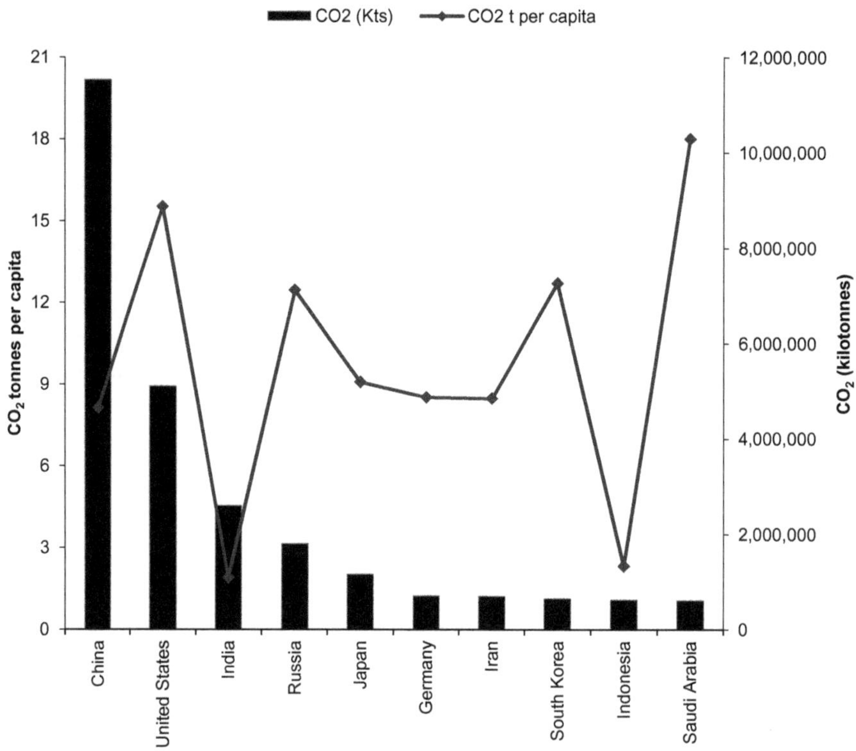

Fig. 3.11 Ranking of CO_2 emissions and CO_2 emissions per capita (2019). *Source* Prepared and adapted by the author from https://datosmacro.expansion.com/

making China the country that has polluted the most with its CO_2 emissions during 2019 in the ranking of countries by CO_2 emissions [51] (Fig. 3.11).

As shown in Fig. 3.11, per capita CO_2 emissions, which have increased in the last year, were 8.12 tonnes per capita in 2019.

Against this backdrop, China has actively participated in the global coalition and in the past six months has made ambitious commitments to control CO_2 emissions. At the UN General Assembly in September 2020, President Xi Jinping announced that China will enhance its Nationally Determined Contribution (NDC) with the goal of reaching peak CO_2 emissions by 2030 and carbon neutrality by 2060 (Fig. 3.12). A key strategy to reduce environmental pollution has been the development of the renewable energy (RE) sector. China's 13th Five-Year Plan reinforces the strategic importance of developing RE as part of the country's energy supply. It provides overall guidance to achieve 15% in energy supply from non-fossil fuel sources by 2020 and 20% by 2030 [79]. In 2020, investment in wind, solar, and hydropower grew by 70.6%, 66.4%, and 19.0% respectively. Newly installed grid-connected wind and solar generation capacity reached a record 71.67 GW and 48.2 GW respectively by the end of 2020 [80].

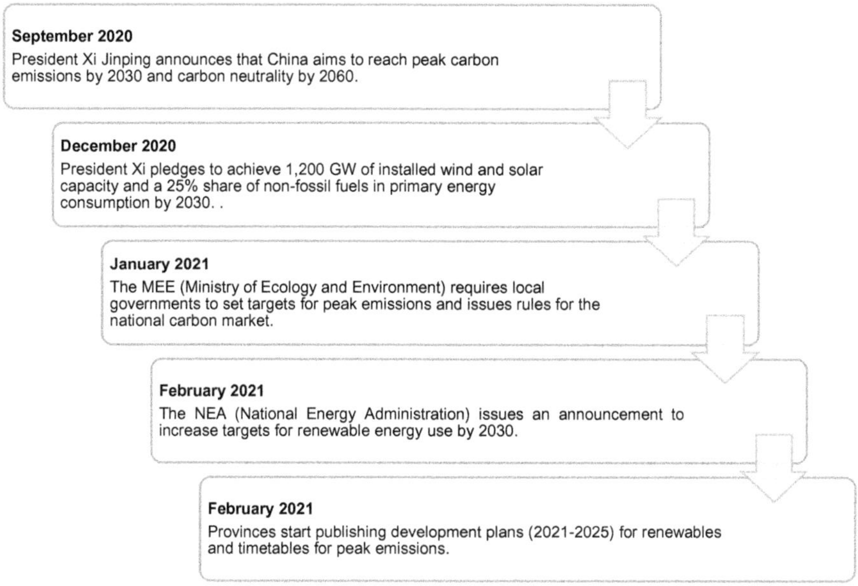

Fig. 3.12 Six months of momentum to net zero. *Source* Prepared and adapted by the author from https://gwec.net/wp-content/uploads/2021/03/GWEC-I-Global-Wind-Report-2021.pdf

The start of China's 14th Five-Year Plan (2021–2025) requires each province to create its own renewable energy development plan and timetable for reaching peak emissions. Jiangsu aims to be the first province to reach peak emissions and targets a cumulative onshore wind capacity of 12GW and offshore wind capacity of 15 GW by 2025.

Guangdong plans to raise the share of non-fossil fuels in primary energy consumption to around 30% by 2025, by which time it targets a cumulative offshore wind capacity of 15 GW. In October 2020, more than 400 companies in the Chinese wind industry adopted the Beijing Declaration, which aims to reach 50 GW of annual installations from 2021 to 2025 and 60 GW from 2026 onwards (Table 3.1).

3.8 Framework of China's Renewable Energy Law and Policy

From the 1990s until 2005, specialised renewable energy policies and laws addressing air pollution in addition to rural energy were gradually issued, the most important of which was the Renewable Energy Law (REL) of 2005. This Act laid the foundations for the development of clean energy in the country, establishing a series of objectives such as diversifying energy supply, protecting the environment, and achieving sustainable development. Subsequently, with the amendment of the aforementioned

Table 3.1 Carbon reduction commitments made by state-owned generators in China

Company	Peaking target	Capacity target
SPIC	Emissions peak by 2023	Clean energy accounts for 60% by 2025 and by 2035
CHN energy	Emissions peak by 2025	70–80 GW renewables added by 2025
Datang	Emissions peak by 2025	Clean energy accounts for 50% by 2025
Huadian	Emissions peak by 2025	75 GW new energy added by 2025, and clean energy accounts for 60% by 2025
Huaneng	Not disclosed yet	80 GW new energy added by 2025, and clean energy accounts for 50% by 2025 and 75% by 2035
CGN (China General Nuclear)	Not disclosed yet	30 GW renewables added by 2025
CTG (China Three Gorges)	Emissions peak by 2023	New energy reaches 70–80 GW by 2025 and carbon neutrality by 2040

Source Prepared and adapted by the author from https://gwec.net/wp-content/uploads/2021/03/GWEC-I-Global-Wind-Report-2021.pdf

"New energy" refers to non-hydro renewable energy sources. *Source* GWEC Market Intelligence, February 2021

law in 2009, a series of tax and investment incentives were also established, allowing for further development and various opportunities for investors [81–83].

In addition to the special laws on renewable energy, other laws also include this issue. For example, the Energy Conservation Law (ECL), which encourages, among other things, the development and use of renewable energies. The Electric Power Act, which supports the generation of electricity from clean energy resources. In addition, environmental and climate protection laws also include renewable energies [84].

China has developed a series of five-year, medium- and long-term plans that set guidelines and targets for the promotion of renewable energy. These national plans enjoy a special position in the system of renewable energy laws and policies and sometimes even produce more effective implementation effects than laws. So far, China has formed a comprehensive renewable energy legal system, essentially based on the REL and complemented by other related laws and policies (Fig. 3.13) [84].

On the other hand, since 2013, the Chinese government has developed a series of renewable energy generation policies, which identify renewable energy as a key component of its development plans [84–88].

A framework containing several policies that can directly support renewable energy generation, or indirectly stimulate its development by limiting fossil energy consumption and increasing the costs of fossil energy generation is defined below as "renewable energy generation policy" (Table 3.2) [89].

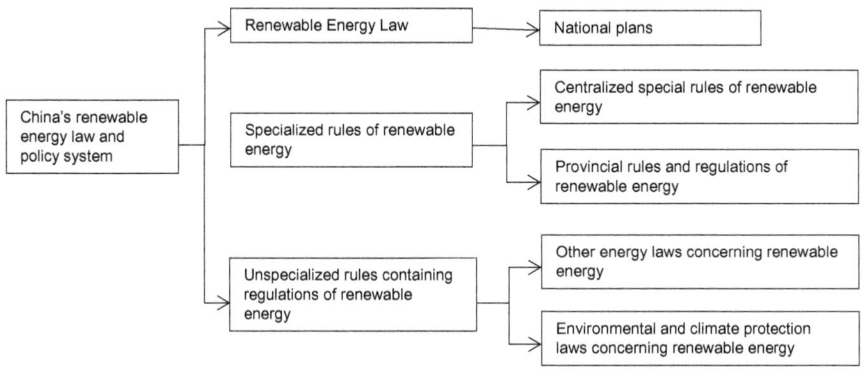

Fig. 3.13 General framework of China's renewable energy law and policy

Table 3.2 Policy framework

Direct policy	Renewable portfolio standards and green certificate trading
	Electricity price subsidy for renewable energy
	Technology R&D subsidy for renewable energy
Indirect policy	Carbon tax
	Eliminate back production capacity
	Optimize industrial structures
	Pollutant discharge permit system

1. Renewable Portfolio Standards (RPS) is a mandatory provision for the share of renewable energy generation in total electricity generation. It was first proposed by the US Environmental Protection Agency (EPA) in the 1990s [90]. Green Certificate Trading is a policy instrument used to implement the renewable portfolio standard. Fossil fuel power generation companies can comply with the RPS by buying green certificates from renewable power generation companies or by paying high penalties. Renewable energy generation companies can earn extra profit from the sale of green certificates.
2. Electricity Price Subsidy for Renewable Energy is a policy to reduce their huge cost disadvantage and maintain the normal operation of renewable energy enterprises. At present, the cost of renewable power generation is about three times higher than the cost of thermal power generation in China, while the benchmark price of renewable power generation is less than twice that of thermal power generation. Therefore, the government has to subsidise the price difference between the reference prices of renewable power generation and thermal power generation.
3. Technology R&D Subsidy for Renewable Energy. The accumulated knowledge and experience, which can help reduce the costs of renewable energy generation,

mainly depends on investment in research and development (R&D). Chinese national and local governments subsidise R&D investments of renewable energy generation companies to some extent to alleviate their financial pressure in developing new technologies.

4. Carbon Tax, as part of the environmental tax, is mainly based on the government's functions to control the negative externality. Carbon tax is a strong policy, which can exert a great impact on the market mechanism and significantly promote renewable energy generation, but also has a great impact on all fields of the national economy.
5. Eliminate Back Production Capacity. The elimination of small backward thermal power plants is an important measure to accelerate the transformation of the power industry, promote energy conservation and emission reduction, and optimise the energy structure. This policy measure will impose additional costs on thermal power generation companies, which will limit their generation in the future.
6. Optimise Industrial Structures. The electricity intensity of the second industry is significantly higher than that of the third industry. Thus, promoting industrial restructuring will reduce electricity demand. In this context, optimising the industrial structure will not only reduce thermal power generation but also increase the share of renewable energy.
7. Pollutant Discharge Permit System. The electricity intensity of the second industry is significantly higher than that of the third industry. Thus, promoting industrial restructuring will reduce electricity demand. In this context, optimising the industrial structure will not only reduce thermal power generation but also increase the share of renewable energy.

In general, the above renewable energy generation policies will either provide direct financial support to the renewable energy generation industry or provide indirect development opportunities by restricting coal consumption and polluting emissions.

3.9 Wind Turbine Installations

The Global Wind Energy Council (GWEC) reports in its Global Wind Report 2021 that 2020 was the best year ever for the global wind industry, showing year-on-year growth of 53%. The 93 GW of new installations in 2020 brings the cumulative global wind capacity to 743 GW [91]. Figure 3.14 shows the world's top five markets in 2020 for new installations: China, the United States, Brazil, the Netherlands, and Germany. These five markets combined accounted for 80.6% of global installations last year, collectively more than 10% higher than in 2019. In terms of cumulative installations, the top five markets at the end of 2020 remained unchanged. Those markets are: China, the US, Germany, India, and Spain, which together account for 73% of the world's total wind installations [92].

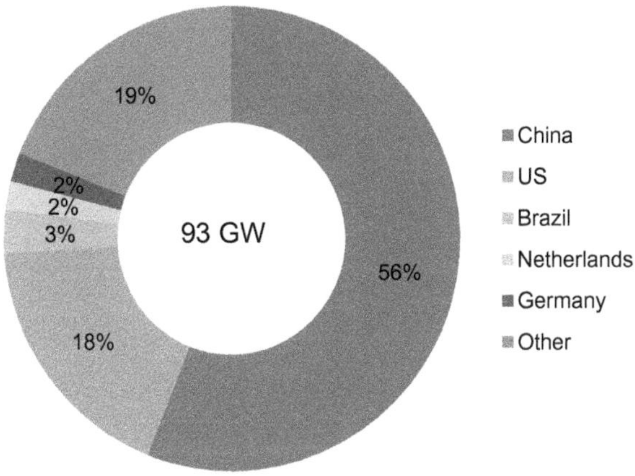

Fig. 3.14 New wind power capacity in 2020 share of top five markets. *Source* Prepared and adapted by the author from https://gwec.net/global-wind-report-2021/

In 2020, 86.9 GW of onshore wind capacity was added globally, an increase of 59% over 2019 and pushing cumulative onshore wind capacity past the 700 GW milestone. This extraordinary increase in 2020 was driven primarily by explosive growth in the world's two largest wind power markets, China and the United States [93]. In China, the National Energy Administration (NEA) reported 68.6 GW of onshore wind installations connected to the grid last year, bringing total onshore installations to more than 272 GW. However, of the 68.6 GW of grid-connected onshore wind, some 26 GW were installed by the end of 2019 but did not actually achieve grid connection until 2020. Excluding this latent volume, new grid-connected installations in China in 2020 were 42.3 GW. Excluding grid connection, new installations were 48.9 GW. The rush to complete onshore wind farms by 2020 was driven by the new policy issued by the National Development and Reform Commission (NDRC) that presented a clear roadmap towards "subsidy-free" onshore wind [93]. This means that projects already approved up to 2018 will continue to receive the feed-in tariff (FiT) if they are connected to the grid before the end of 2020. From 1 January 2021, all new approved onshore wind projects must reach grid parity (currently based on the regulated feed-in tariff for coal power in each province) [94] (Fig. 3.15).

In the offshore market, 6.1 GW were commissioned worldwide last year, making 2020 the second best year ever. China led the world in annual new offshore wind installations for the third year in a row, with more than 3 GW of new offshore wind capacity in 2020 [93, 95–97] (Figs. 3.16 and 3.17).

As mentioned above and under China's 14th Five-Year Plan (2021–2025), Jiangsu aims to be the first province to reach peak emissions and targets a cumulative onshore wind capacity of 12GW and offshore wind capacity of 15 GW by 2025. Guangdong plans to raise the share of non-fossil fuels in primary energy consumption to around 30% by 2025, by which time it targets a cumulative offshore wind capacity of 15 GW. Last autumn, China announced a plan to peak emissions by 2030 and achieve carbon neutrality by 2060. To back up this commitment, the wind industry has published the "Beijing Declaration on Wind Energy" with the ambition to install 3000 GW of wind power by 2060. This will require an average annual installation of 50 GW over

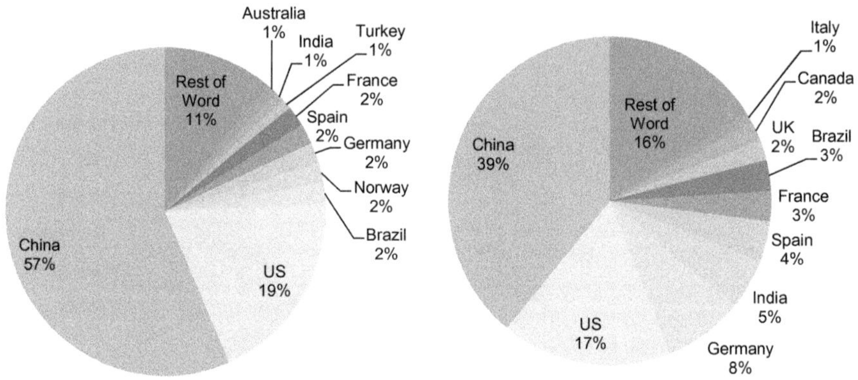

Fig. 3.15 New wind energy installations (left) and total installations (right) onshore. *Source* Prepared and adapted by the author from https://gwec.net/global-wind-report-2021/

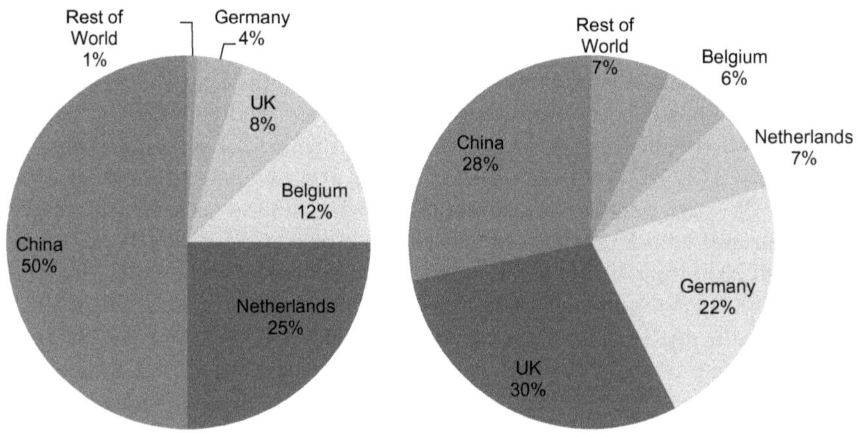

Fig. 3.16 New wind installations (left) and total installations (right) offshore. *Source* Prepared and adapted by the author from https://gwec.net/global-wind-report-2021/

the next five years (2021–2025) and at least 60 GW per year after 2025. This would bring China's cumulative wind capacity to 800 GW by 2030 and 3000 GW by 2060. The tremendous growth showcase of China's wind industry in 2020 shows that 50 GW per year is not only possible but would reinforce the country's progress towards its goal of reaching peak emissions by 2030 and ensure a cost trajectory towards carbon neutrality by 2060 [93, 98] (Fig. 3.18).

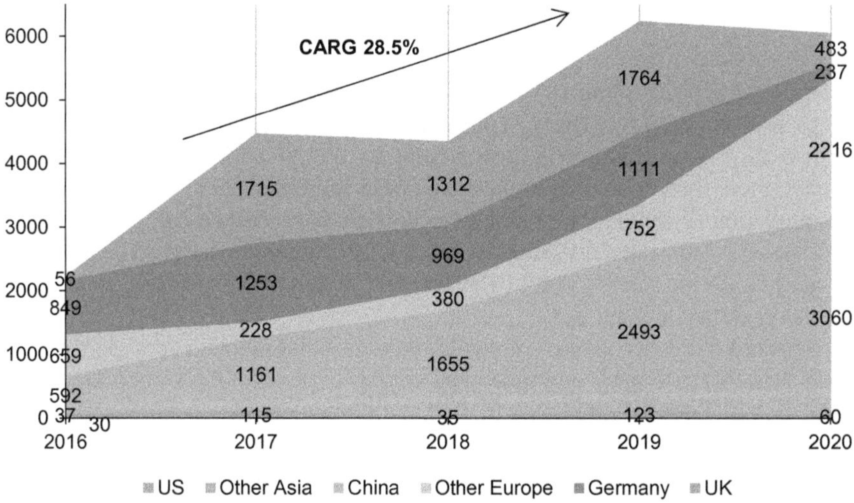

Fig. 3.17 New offshore installation (MW). *Source* Prepared and adapted by the author from https://gwec.net/global-wind-report-2021/

Net-zero target, if any

- Net zero carbon emissions by 2060

Status of the legislation

- Medium-term targets under the net-zero goal will be formulated in Five-Year Plans, including 14th Five-Year Plan (2021-2025)

Public investment announced alongside the net-zero target

- No such announcement available yet

NDC, as of February 2021

- Lowering carbon intensity by more than 65% by 2030 from 2005 levels

Renewable energy targets

- Reduce share of non-fossil fuels in primary energy consumption to around 25% by 2030
- Increase installed capacity of wind and solar power to more than 1,200GW by 2030.

Installed wind capacity as of end of 2020

- 272GW onshore and 9GW offshore

Key technology strategy on energy transition

- Wind and solar power will take a leading role, with nuclear and hydropower as subsidiary elements
- Innovative grid system
- Storage, hydrogen, CCS technologies to scale up

Other drivers of clean energy transition

- National carbon trading market will be established in 2021
- Promoting industrial restructuring
- Improving energy efficiency (industry, building, transportation, public institutions)
- Establishing market mechanisms (pricing, taxes, financial) for low-carbon development
- Increasing carbon sequestration capacity

Fig. 3.18 Overview of China's net zero plans. *Source* Prepared and adapted by the author from https://files.wri.org/d8/s3fs-public/VISION2050_English_Summary_Corrected.pdf

3.10 Solar PV Installations

Currently, solar photovoltaic (PV) energy is playing an important role in many countries to displace fossil fuel energy. China has great opportunities to use solar energy due to its huge solar resource [99]. According to the China Meteorological Administration; more than two-thirds of the country has more than 2000 h of sunshine per year, providing an annual solar radiation equivalent of more than 5.02 × 106 kJ/m^2. The total solar radiant energy potential is 1.7×10^{12} tonnes of standard coal equivalent per year for the country [100, 101]. The distribution of the solar energy resource in China is shown in Table 3.3.

Compared to other countries of similar latitude, solar radiant energy in China is higher than that in Europe and Japan and similar to those in the United States. As can be seen in Table 3.3, provinces located at different latitudes and longitudes have different levels of solar irradiance. The country can be divided into five different regions, from I to V [102]. The distribution of China's solar energy resources in different areas varies significantly; In general, the solar resources in the western region (such as Ningxia, Gansu, Xinjiang, Qinghai, and Tibet) are larger than those in the eastern region (such as Guangdong, Shaanxi, Anhui, Heilongjiang, Zhejiang, Fujian, Hunan, and Hubei), and the resources of the northern region (such as Hebei, Shanxi, Inner Mongolia, Shandong Hena, Jilin, Liaoning, and Shaanxi) are higher than those of the southern region (such as Sichuan, Guizhou, Chongqing, Guangxi, and Jiangxi).

China began generating solar photovoltaic (PV) power in the 1960s, and power generation is the dominant form of solar energy [103, 104]. After a long period of development and due to China's policy, its solar PV industry has made spectacular and unprecedented progress in the last 10 years [105]. On the manufacturing side, China has created a complete PV industry chain with international competitiveness. In 2017, the production of the entire industrial chain in China accounted for 70% of global production (Fig. 3.19). The module production capacity is about 120 GW. The industry employs 2.5 million people. The annual production value of the entire industry chain reaches 500 billion yuan [106].

The average annual growth rate of cumulative installed capacity of solar PV was 122% during the Twelfth Five-Year Plan period. Figure 3.20 shows China's new and cumulative installed capacity from 2000 to 2020. In 2000, the installed capacity was only 3 MW. In 2007, PV module production reached 1.1 GW, and 11 PV companies were listed abroad, making the country the largest PV producer. However, the domestic PV market was only 20 MW, which meant that up to 98% of PV modules were exported. At the end of 2008, the Chinese government held tenders for PV plant concessions for the first time and in 2009 launched the photoelectric building project and Golden Sun Demonstration Project, thus kick-starting the domestic PV market [107, 108]. From 2008 to 2017, the PV market grew at an average annual rate of 146%, and the export of PV modules increased from 98 to 34%, making it the largest solar PV market in the world. New installed capacity has been number one for 5 consecutive years since 2005, and cumulative capacity number 1 for 3

Table 3.3 Solar energy resources in different regions of China

Type	Annual sunshine hours (h/a)	Total annual solar radiation (MJ/m^{2}a)	Total annual solar energy per m^2 expressed in units of energy produced by kg of standard coal (kg)	Provinces
I	3200–3300	6680–8400	225–285	Northern Ningxia, Northern Gansu, Southeastern Xinjiang, Western Qinghai, and Western Tibet
II	3000–3200	5852–6680	200–225	Northwestern Hebei, Northern Shanxi, Southern Inner Mongolia, Southern Ningxia, Central Gansu, Eastern Qinghai, Southeastern Tibet, and Southern Xinjiang
III	2200–3000	5016–5852	170–200	Southeastern Shandong, Southeastern Henan, Northwestern Hebei, Southern Shanxi, Northern Xinjiang, Jilin, Liaoning, Yunnan, Northern Shaanxi, Southeastern Gansu, Southern Guangdong, Southern Fujian, Northern Jiangsu, Northern Anhui, Tianjin, Beijing, and Southwestern Taiwan
IV	1400–2000	4190–5016	140–170	Hunan, Hubei, Guangxi, Jiangxi, Zhejiang, Northern Fujian, Northern Guangdong, Southern Shaanxi, Southern Anhui, Heilongjiang, and Northeastern Taiwan

(continued)

Table 3.3 (continued)

Type	Annual sunshine hours (h/a)	Total annual solar radiation (MJ/m^{2}a)	Total annual solar energy per m^2 expressed in units of energy produced by kg of standard coal (kg)	Provinces
V	1000–1400	3344–4190	115–140	Sichuan, Guizhou, and ChongqingZ.-y. Zhao et al./Renewable and Sustainable Energy Reviews 21 (2013) 229–236,230

Source Prepared and adapted by the author from /http://www.ck365.cn/anli/16148.htmlS

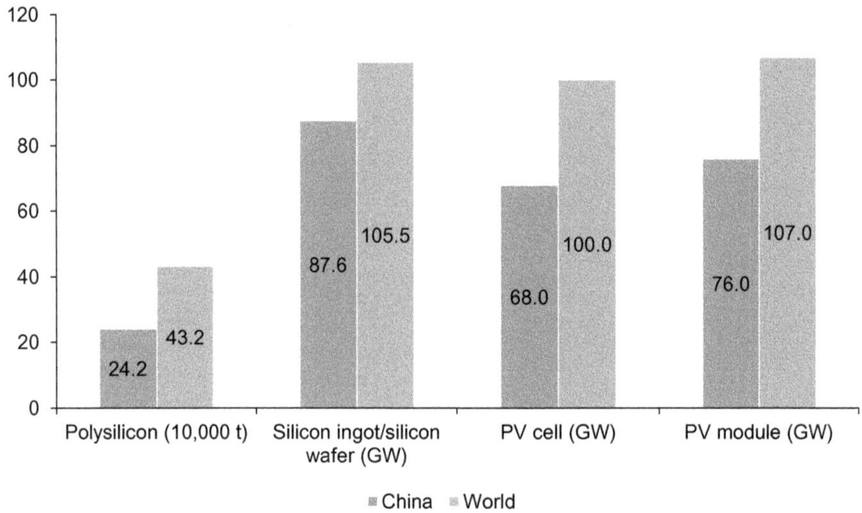

Fig. 3.19 Output Chinese PV industry (2017). *Source* Prepared and adapted by the author from http://www.chinapv.org.cn/annualreport/584.html

consecutive years since 2015. In 2020 these values were 40 GW and 253.43 GW in 2020, respectively [99].

3.11 Solar Thermal Installations

The solar thermal industry in China has developed rapidly since 1995 [109]. In 1998, 3.4 million m^2 of solar collectors were produced, while in 2002, 10 million m^2 were produced, corresponding to an annual growth rate of about 31%. China's total

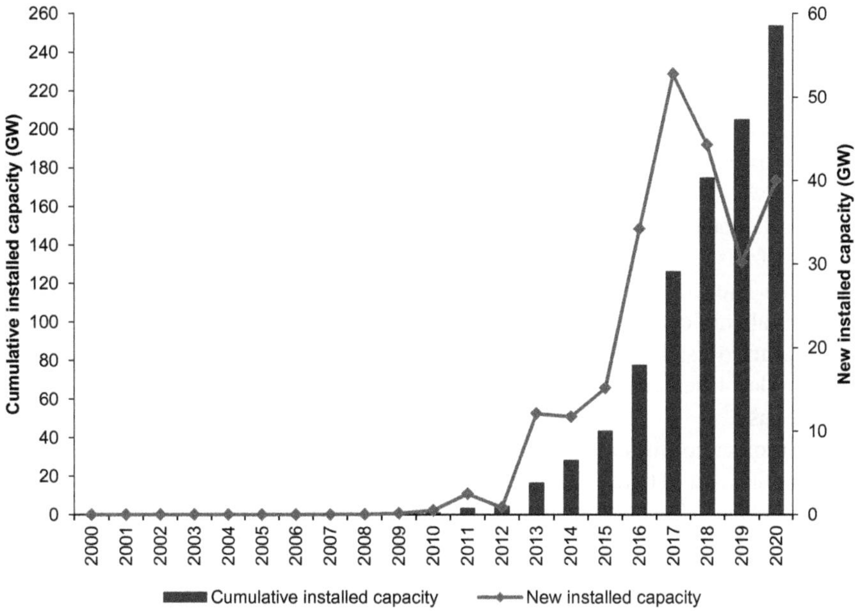

Fig. 3.20 China's new and cumulative installed capacity from 2000 to 2020. *Source* Prepared and adapted by the author from https://www.statista.com/statistics/673375/newly-installed-capacity-of-solar-power-in-china/

installed solar thermal capacity (cumulative total) has accounted for about 70% of all global capacity since 2013 [110], with a significant difference of more than ten times the capacity of the second largest country.

The reasons why solar hot water systems are successful in China can be summarised as follows:

- The central government is proactively promoting their application [111, 112];
- China's economy is developing rapidly and its standard of living is rising, especially in rural areas, and more and more households need hot water 24 h a day;
- China's solar thermal industry consisted of a complete industrial chain from raw material processing to manufacturing, sales, and service;
- China's central government and local governments promote solar hot water systems with subsidies [112].

There are even policies of mandatory installation of solar water heaters for urban residential buildings of less than 12 stories, which were issued by some local governments in recent years.

Although China is the largest producer of solar thermal collectors in the world, the installed capacity of solar hot water systems per capita in China is still quite low compared to other countries, which means that there is huge market potential in China.

3.11.1 Solar Heating in Tibet at Langkazi (China)

Tibet has the most abundant solar energy resources in the world, with annual global radiation on the horizontal surface exceeding 2000 kWh/m^2/year. Conventional energy infrastructures, such as natural gas pipelines, are very expensive. The Chinese central government attaches great importance to improving living standards in Tibet and has allocated funds to build heating facilities for the Tibetan people. The town of Langkazi is located in a very cold and high-altitude area, where heating is very necessary. The elevation is 4200 m and the boiling temperature of the water is 84 °C. The central and local governments started building a large SWH plant in 2017. The project aims to test the reliability and applicability of the deployed system and provide an example of best practices to solve the problem of heating problems in Tibetan urban areas.

The project consists of a set of collectors, a buffer storage, an electrical boiler system, hot water network, and indoor heating terminal units. The project is divided into two phases. The first phase shows a collector area of 22,275 m^2, which was completed in November 2018. The inclination of the solar collectors is 40°. The size of the water well storage is 15,000 m^3.

The storage in the water well is used as short-term storage, not as seasonal storage, as the storage losses would be too high for cost-effective seasonal storage at this altitude. The solar heat can be delivered directly to the hot water network or to the storage. Two electric boilers (2×1.5 MW) were installed as backup heat sources. In the first phase, 82,600 m^2 of building area with 26 communities were connected to the district heating (DH) network. In the second phase, 14,525 m^2 of collector surface area will be added when more consumers are connected to the DH network in the near future. The collector array is located about 2 km south of the city of Langkazi.

The designed heating period is from 23 September to 31 May (251 days in total). The design outdoor temperature during the heating period is −14.4 °C. The indoor design temperature is 18 °C. The solar yield is higher in winter than in summer, because there is a lot of rain in summer and the irradiation is lower. The DH network only provides heating and not hot water. Like many DH grids in China, it is switched off during the summer months. In the future, hot water could be added to increase the summer load. The solar collector field aims to cover more than 90% of the heat demand.

Langkazi Tibet Solar Heating is a 100% subsidy project. The project received a subsidy of 175 million RMB from the central government. It was realised as a turnkey solution by a consortium of Beijing Heating Project Design Co. Ltd, Suneast Arcon-Sunmark, and Sunrain Group.

3.12 Conclusions

The demand for energy in China has increased rapidly and has reached an unprecedented level due to the country's rapid economic growth and modern development. Energy shortages and the deterioration of the natural environment have seriously constrained the sustainable development of China's economy and society.

The year 2021 marks the beginning of China's 14th Five-Year Plan. With the new situations at home and abroad, China faces both opportunities and challenges. Taking into account various factors such as power substitution, low base number in the previous year, as well as the uncertainty caused by the pandemic and the changing global environment, China's total energy demand is expected to grow at a faster rate in the first half of 2021 and slow down thereafter, with annual growth of 6–7% throughout the year [90].

These real demands need to be urgently addressed by the Renewable Energy Law (REL) and other related laws and policies. China's current legal framework for renewable energy is of little help due to its fragmentation, obsolescence, and poor operability. The Chinese government needs to integrate the system of renewable energy laws and policies, improve the viability of renewable energy, apply market economics, provide and rationally plan renewable energy generation. It is hoped that through these means, it can provide more systematic and efficient support for the overall healthy development of the renewable energy industry.

Renewable energy is a solution that strikes a balance between energy security, environmental protection, and economic efficiency. It is undeniable that renewable energy in China is in a phase of rapid development. However, compared to its potential, it is still under-exploited. There is still much scope for developing large-scale renewable sources in the future energy system, along with technological and economic development.

References

1. Chen SY (2013) Energy, environment and economic transformation in China. Routledge, Taylor & Francis Group, London and New York
2. The Oxford Institute for Energy Studies (2018) The outlook for natural gas and LNG in China in the war against air pollution, Dec 2018, pages 11–20
3. Song C, Wu L, Xie Y, He J, Chen X, Wang T et al (2017) Air pollution in China: status and spatiotemporal variations. Environ Pollut 227:334–347
4. Kuhn B (2019) Ecological civilisation in China. Dialogue Civilizations Research Institute, en ligne.
5. Guo B, Wang Y, Zhang X, Che H, Zhong J, Chu Y, Cheng L (2020) Temporal and spatial variations of haze and fog and the characteristics of PM2.5 during heavy pollution episodes in China from 2013 to 2018. Atmos Pollut Res 11(10):1847–1856
6. Poon JP, Casas I, He C (2006) The impact of energy, transport, and trade on air pollution in China. Eurasian Geogr Econ 47(5):568–584

7. Díaz G, Andrea D (2020) El cambio climático y la respuesta de las grandes potencias. El caso de Estados Unidos y China. Análisis Político 33(99):121–142. https://doi.org/10.15446/anpol.v33n99.90987
8. EIA (2020) Independent statistics & analysis. U.S. Energy Information Administration. Country Analysis Executive Summary: China. Last Update: September 30,2020. https://www.eia.gov/international/content/analysis/countries_long/China/china.pdf. Accessed on 08 Dec 2020
9. Council, G. W. E. (2017). GWEC| GLOBAL WIND REPORT 2021. https://gwec.net/global-wind-report-2021/ [accessed on May 14, 2020].
10. Zhu X, Zhang P, Wei Y, Li Y, Zhao H (2019) Measuring the efficiency and driving factors of urban land use based on the DEA method and the PLS-SEM model—a case study of 35 large and medium-sized cities in China. Sustain Cities Soc 50:101646
11. Tregear TR (2017) A geography of China. [Illustr.]. Transaction Publishers
12. Zhang J, Wang Y, Zhang B, Zhao H (2015) Evolution of the NE Qinghai-Tibetan Plateau, constrained by the apatite fission track ages of the mountain ranges around the Xining Basin in NW China. J Asian Earth Sci 97:10–23
13. Wang X-P, Wang Z-H, Fang J-Y (2004) Mountain ranges and peaks in China. Biodivers Sci 12(1):206
14. Zhang X, Li MJ, Zhan LQ, Wu W, Liu HB (2020) Boron availability in top-and sub-soils as affected by topography and climate. Nutr Cycl Agroecosyst 118(1):91–101
15. Wei K, Lin R (1994) The influence of the monsoon climate on the isotopic composition of precipitation in China. Geochimica 23(1):33–41
16. Zhao E, Zhao G, Ju J, Zhao S (2011) Impacts of activity of tropical monsoon air mass on the rainy season of Yunnan. J Yunnan Univ Nat Sci Ed 33(4):428–438
17. Domrös M, Peng G (2012) The climate of China. Springer Science & Business Media
18. Tu CW, Lu X (1938) The air masses of China. Acta Meteorol Sin 5:175–218
19. Lee KS, Grundstein AJ, Wenner DB, Choi MS, Woo NC, Lee DH (2003) Climatic controls on the stable isotopic composition of precipitation in Northeast Asia. Clim Res 23(2):137–148
20. Yu Y, Cai M, Ren R, Van Den Dool HM (2015) Relationship between warm airmass transport into the upper polar atmosphere and cold air outbreaks in winter. J Atmos Sci 72(1):349–368
21. Zhang J, Zhang M, Li Y, Qin J, Wei K, Song L (2020) Analysis of wind characteristics and wind energy potential in complex mountainous region in southwest China. J Cleaner Product 274:123036
22. Wang HJ, Sun JQ, Chen HP, Zhu YL, Zhang Y, Jiang DB et al (2012) Extreme climate in China: facts, simulation and projection. Meteorol Z 21(3):279
23. Shaowu W, Jinhong Z, Jingning C (2004) Interdecadal variability of temperature and precipitation in China since 1880. Adv Atmos Sci 21(3):307–313
24. Yang B, Braeuning A, Johnson KR, Yafeng S (2002) General characteristics of temperature variation in China during the last two millennia. Geophys Res Lett 29(9):38–41
25. Zhai P, Zhang X, Wan H, Pan X (2005) Trends in total precipitation and frequency of daily precipitation extremes over China. J Clim 18(7):1096–1108
26. Liu B, Xu M, Henderson M, Qi Y (2005) Observed trends of precipitation amount, frequency, and intensity in China, 1960–2000. J Geophys Res Atmos 110(D8)
27. Liang L, Li L, Liu Q (2011) Precipitation variability in Northeast China from 1961 to 2008. J Hydrol 404(1–2):67–76
28. Yang F, Lau KM (2004) Trend and variability of China precipitation in spring and summer: linkage to sea-surface temperatures. Int J Climatol J Royal Meteorol Soc 24(13):1625–1644
29. Chen B, Xu XD, Yang S, Zhang W (2012) On the origin and destination of atmospheric moisture and air mass over the Tibetan Plateau. Theoret Appl Climatol 110(3):423–435
30. The World Bank, Source: Global Solar Atlas 2.0 (2019) Solar resource data: Solargis. https://solargis.com/maps-and-gis-data/download/china. Accessed on 14 Apr 2021
31. Bachman D, Bachman DM (2006) Bureaucracy, economy, and leadership in China: the institutional origins of the great leap forward. Cambridge University Press

32. Shen Z, Xia Y (2011) The great leap forward, the people's commune and the Sino-Soviet split. J Contemp China 20(72):861–880
33. Domenach JL, Selden M (2019) The origins of the great leap forward: the case of one Chinese province. Routledge
34. Karl RE (2010) Mao Zedong and China in the twentieth-century world: a concise history. Duke University Press
35. Knight N (2002) Mao Zedong's "Sinification of Marxism" 1985. In: Mao Zedong and China's Revolutions. Palgrave Macmillan, New York, pp 197–204
36. Yang, Z. (2016). Hu Yao-Bang: A Chinese Biography: A Chinese Biography. Routledge.
37. Castrillón D, Álvarez L (2019) Elementos del confucianismo en el desarrollo de la economía china, 1949–2012. Papel Político 24(2)
38. Béja JP (ed) (2010) The impact of China's 1989 Tiananmen massacre, vol 17. Routledge
39. Baum R (2012) The impact of China's 1989 Tiananmen Massacre
40. Jin Q (1999) The culture of power: the Lin Biao incident in the cultural revolution. Stanford University Press
41. Lau WK (1999) The 15th congress of the Chinese communist party: milestone in China's privatization. Cap Class 23(2):51–87
42. Mavroidis PC, Sapir A (2019) China and the world trade organisation: towards a better fit. Bruegel Working Papers
43. Womack B (2017) International Crises and China's rise: comparing the 2008 global financial crisis and the 2017 global political crisis. Chin J Int Polit 10(4):383–401
44. Zhang M (2009) The impact of the global crisis on China and its reaction. Elcano Newslett 55:7
45. Prashantham S, Yip GS (2016) Engaging with startups in emerging markets
46. Men LR, Chen ZF, Ji YG (2018) Walking the talk: an exploratory examination of executive leadership communication at startups in China. J Public Relations Res 30(1–2):35–56
47. Li L (2018) China's manufacturing locus in 2025: With a comparison of "Made-in-China 2025" and "Industry 4.0." Technol Forecast Soc Change 135:66–74
48. Wübbeke J, Meissner M, Zenglein MJ, Ives J, Conrad B (2016) Made in China 2025. Mercator Institute for China Studies. Papers on China 2:74
49. Zenglein MJ, Holzmann A (2019) Evolving made in China 2025. MERICS Papers China 8:78
50. Datos Macro (2020) https://datosmacro.expansion.com/. Accessed on 11 Nov 2020
51. Feenstra RC, Wei SJ (eds) (2019) China's growing role in world trade. University of Chicago Press
52. World Bank (2020) GDP growth (Annual %). Available online: https://data.worldbank.org/indicator/NY.GDP.MKTP.KD.ZG?locations=CN. Accessed on 14 Dec 2020
53. Huang Y, Pagano M, Panizza U (2020) Local crowding-out in China. J Financ 75(6):2855–2898
54. Chen Y, Wang Z, Zhong Z (2019) CO_2 emissions, economic growth, renewable and non-renewable energy production and foreign trade in China. Renew Energy 131:208–216
55. Xu L, Chen N, Chen Z (2017) Will China make a difference in its carbon intensity reduction targets by 2020 and 2030? Appl Energy 203:874–882
56. World Bank. Available online: https://datos.bancomundial.org/pais/china?view=chart. Accessed on14 Dec 2020
57. Zhao J, Tang J (2018) Understanding agricultural growth in China: an international perspective. Struct Chang Econ Dyn 46:43–51
58. Mrówczyńska-Kamińska A, Bajan B (2019) Importance and share of agribusiness in the Chinese economy (2000–2014). Heliyon 5(11):e02884
59. Wu Y (2007) Service sector growth in China and India: a comparison. China Int J 5(1):137–154
60. Chow GC (2017) Capital formation and economic growth in China, pp 1186–1221. Brill
61. Jiang Y, Chen X, Valdmanis V, Baležentis T (2019) Evaluating economic and environmental performance of the Chinese industry sector. Sustainability 11(23):6804
62. National Bureau of Statistics of China. China statistical yearbook. http://www.stats.gov.cn/tjsj/ndsj/2019/indexeh.htm. Accessed on 14 Mar 2021

63. Dudley B (2018) BP statistical review of world energy. BP Statistical Review, London, UK. Accessed on 06 Aug 2020
64. IMF World Economic Outlook Update January 2019, https://www.imf.org/en/Publications/WEO/Issues/2019/01/11/weo-update-january-2019. Accessed on 23 Mar 2021
65. IMF (2021) Questions and answers on the IMF's $50 billion rapid-disbursing emergency financing facilities. Mar 19, 2020. https://www.imf.org/en/About/FAQ/50-billion-rapiddisbursing-emergency-financing-facilities. Accessed on 23 Mar 2021
66. China Daily, "Significant coal-to-liquid project in production in Ningxia," December 28, 2016.
67. EIA (2021) World energy outlook 2017: China, Nov 2017. https://www.iea.org/reports/world-energy-outlook-2017-china. Accessed on 23 Mar 2021
68. CNOOC Limited, 2019 Annual Results, Mar 25, 2020, page 18.
69. Reuters (2019) China CNOOC to start pumping at large deepsea gas field at end-2021: official, Sept 24, 2019; CNOOC Limited, 2019 Mid-Year Review Analyst Briefing, Aug 29, 2019, pages 9 and 27
70. U.S. Energy Information Administration estimates using growth rates from National Bureau of Statistics of China, "Statistical Communiqué of the People's Republic of China on the 2018 National Economic and Social Development," February 28, 2019.
71. Reuters (2019) China boosts coal mining capacity despite climate pledges, Mar 26, 2019. International Energy Agency, Coal 2019, pages 32–33; International Energy Agency, Coal 2018ss, pages 32–33
72. International Energy Agency, Coal 2019, pages 31, 110, 123–124; Xinhua News, Haoji Railway for coal transportation opens to traffic. Sept 29, 2019
73. CEC (2020) Analysis and forecast of China Power Demand-Supply Situation 2020–2021. Pubdate: 2021–02–09 04:34:18. https://english.cec.org.cn/detail/index.html?3-1128. Accessed on 07 June 2020
74. S&P Global Platts (2019) Analysis: China's rationalized coal-to-gas policy crimps winter gas demand growth, Nov 7, 2019; South China Morning Post, "China's rush to dump coal for gas is modified to avoid repeat of winter heating crisis, Aug 29, 2018. The Oxford Institute for Energy Studies, China's Natural Gas Development Report—Reality Check, Dec 2019, page 1
75. International Energy Agency, Gas Market Report 2019, page 20; The Oxford Institute for Energy Studies, The outlook for natural gas and LNG in China in the war against air pollution, Dec 2018, page 33; FACTS Global Energy, East of Suez Gas Databook: China Natural Gas Outlook, September 2019, page 17
76. FACTS Global Energy, East of Suez Gas Databook: China Natural Gas Outlook, Sept 2019, page 28
77. FACTS Global Energy, China Gas Monthly, March 18, 2020, pages 3–4; S&P Global Platts, China's Jan-Feb natural gas import growth slows to 2.8% in 2020 from 18.5% in 2019, Mar 7, 2020
78. International Energy Agency, Coal 2019, pages 16–18
79. NDRC (2016) Renewable energy development, the 13th five-year plan. National Development and Reform Commission. http://ghs.ndrc.gov.cn/ghwb/gjwngh/201605/P020160516532440059919.pdf. 3, 2016. Accessed on 04 May 2021
80. Shen W (2017) Who drives China's renewable energy policies? Understanding the role of industrial corporations. Environ Dev 21:87–97
81. Schuman S, Lin A (2012) China's renewable energy law and its impact on renewable power in China: progress, challenges and recommendations for improving implementation. Energy Policy 51:89–109
82. Liu J (2019) China's renewable energy law and policy: a critical review. Renew Sustain Energy Rev 99:212–219
83. Huo ML, Zhang DW (2012) Lessons from photovoltaic policies in China for future development. Energy Policy 51:38–45
84. Mischke P, Karlsson KB (2014) Modelling tools to evaluate China's future energy system–a review of the Chinese perspective. Energy 69:132–143

85. Wang Y, Zhou S, Huo H (2014) Cost and CO_2 reductions of solar photovoltaic power generation in China: perspectives for 2020. Renew Sustain Energy Rev 39:370–380
86. Guo X, Guo X (2015) China's photovoltaic power development under policy incentives: a system dynamics analysis. Energy 93:589–598
87. Guo X, Guo X (2016) Nuclear power development in China after the restart of new nuclear construction and approval: a system dynamics analysis. Renew Sustain Energy Rev 57:999–1007
88. Liu L, Zong H, Zhao E, Chen C, Wang J (2014) Can China realize its carbon emission reduction goal in 2020: from the perspective of thermal power development. Appl Energy 124:199–212
89. Liu D, Liu M, Xu E, Pang B, Guo X, Xiao B, Niu D (2018) Comprehensive effectiveness assessment of renewable energy generation policy: a partial equilibrium analysis in China. Energy Policy 115:330–341
90. Costs and benefits of renewables portfolio standards in the United States B. Galen, B. Lori, H. Jenny, F. Francisco and W. Ryan Renew Sustain Energy Rev 52:523–533
91. https://webcache.googleusercontent.com/search?q=cache:SJo8SyYNV5cJ:https://www.windpowermonthly.com/article/1704954/offshore-wind-spending-reaches-record-high-2020+&cd=1&hl=en&ct=clnk&gl=uk. Accessed on 01 Mar 2020
92. Lee J, Zhao F (2020) GWEC|global wind report 2021. https://gwec.net/global-wind-report-2021/. Accessed on 01 Mar 2020
93. https://en.ndrc.gov.cn/. Accessed on 01 Mar 2020
94. Du Y, Takeuchi K (2020) Does a small difference make a difference? Impact of feed-in tariff on renewable power generation in China. Energy Econ 87:104710
95. Chen J (2011) Development of offshore wind power in China. Renew Sustain Energy Rev 15(9):5013–5020
96. Esteban MD, Diez JJ, López JS, Negro V (2011) Why offshore wind energy? Renew Energy 36(2):444–450
97. Díaz, H., & Soares, C. G. (2020). Review of the current status, technology and future trends of offshore wind farms. Ocean Engineering, 209, 107381.
98. WRI China (2020) Accelerating the net-zero transition: strategic action for China's 14th five-year plan. https://files.wri.org/d8/s3fs-public/VISION2050_English_Summary_Corrected.pdf. Accessed on 01 Mar 2020
99. Zhao ZY, Zhang SY, Hubbard B, Yao X (2013) The emergence of the solar photovoltaic power industry in China. Renew Sustain Energy Rev 21:229–236
100. Peidong Z, Yanli Y, Yonghong Z, Lisheng W, Xinrong L (2009) Opportunities and challenges for renewable energy policy in China. Renew Sustain Energy Rev 13(2):439–449
101. Luo YJ, He XN, Wang CG (2005) Application technology of solar energy. Chemistry and Industry Press
102. The distribution features of solar energy resources in China (2020) http://www.ck365.cn/anli/16148.htmlS. Accessed on 01 July 2020
103. Wang Q (2010) Effective policies for renewable energy—the example of China's wind power—lessons for China's photovoltaic power. Renew Sustain Energy Rev 14(2):702–712
104. Xu M, Xie P, Xie BC (2020) Study of China's optimal solar photovoltaic power development path to 2050. Resour Policy 65:101541
105. Sun B, Yu Y, Qin C (2017) Should China focus on the distributed development of wind and solar photovoltaic power generation? A comparative study. Appl Energy 185:421–439
106. Wang S (2020) Current status of PV in China and its future forecast. CSEE J Power Energy Syst 6(1):72–82
107. Wang Y, Luo G, Kang H (2017) Successes and failures of china's golden-sun program. In: 2017 6th international conference on energy, environment and sustainable development (ICEESD 2017). Atlantis Press
108. Zhiqiang Y (2005) Development of solar thermal systems in China. Sol Energy Mater Sol Cells 86(3):427–442

109. International Energy Agency (2017) IEA solar heating & cooling programme. https://www.iea-shc.org/. 2017 [Online]. Available: Mar 2017
110. Xie H, Zhang C, Hao B, Liu S, Zou K (2012) Review of solar obligations in China. Renew Sustain Energy Rev 16(1):113–122
111. Huang J, Tian Z, Fan J (2019) A comprehensive analysis on development and transition of the solar thermal market in China with more than 70% market share worldwide. Energy 174:611–624
112. Goess S, De Jong M, Ravesteijn W (2015) What makes renewable energy successful in China? The case of the Shandong province solar water heater innovation system. Energy Policy 86:684–696

Chapter 4
Wind Energy and Solar PV Developments in the EU

4.1 Introduction

The European Union (EU) is a single political and economic union composed of 27 European countries. The organization that was to become the EU, the European Economic Community, was created in 1958, as a result of the Second World War, with the objective of increasing economic cooperation between its six members: Belgium, France, Germany, Italy, Luxembourg, and the Netherlands. Its first phases were to promote economic cooperation between the members in the belief that trade produces interdependence between countries and it reduces the potential for conflict.

Since that time, 22 more countries have joined (with the UK leaving the EU in 2020) and a huge single market (also known as the 'internal' market) has been created and continues to move towards realising its full potential. Although at first, the EU was a purely economic union, it has grown into an organization that covers many different policy areas from external relations and security, justice, and migration to climate, environment, and health. As a reflection of this development, in 1993 this organizations name changed from the European Economic Community to the European Union.

The EU has achieved more than half a century of peace, stability, and prosperity. It has improved the living standards and has created a single European currency, the euro. Currently, the euro is used in 19 countries with more than 340 million EU citizens.

The EU is the world's largest economy, accounting for more than 20% of the world's Gross Domestic Product (GDP). Thanks to the size of its GDP (around 14 trillion euro) and the openness of its market, with exports amounting to 3142 billion euro and imports of goods and services amounting to 2842 billion euro, the EU has played a key role in influencing the international trading system, especially by supporting the World Trade Organization (WTO). Economic openness has brought significant benefits to the EU and will continue to do so since more than 30 million jobs in the EU depend on external trade and over the next 15 years, 90% of world economic growth is expected to be generated outside Europe.

T. Muneer et al., *The Coming of Age of Solar and Wind Power*, Green Energy and Technology, https://doi.org/10.1007/978-3-030-92010-4_4

The challenges related to the energy area that the European Union is facing include its growing dependence on imports, the increasing global energy demand, climate change, and the need to increase the share of renewable energy. The European energy policy essentially consists of a series of measures to face the previously mentioned challenges.

The current energy situation depends strongly on non-renewable energy sources. Even though the EU is a pioneer in renewable energy, with high efforts in its legislation to better integrate renewable sources into European energy systems, one of its biggest problems is its energy dependence. Western Europe is totally lacking in non-renewable energy sources. It once had large deposits of coal, but it lost importance and interest with the advent of natural gas, petroleum, and nuclear energy. The diversification of its energy sources is also an important aim to remedy the lack of self-energy sources.

The 58% of the energy consumption in the EU comes from imported energy. In 2018, petroleum products consisted of two-thirds of the imported energy, followed by natural gas (24%) and solid fossil fuels (8%).

Regarding renewable energy sources, in the first half of 2020, for the first time, the EU generated more electricity from renewable sources than from fossil fuels. It is also important to mention that the renewable energy share increased from 9.6% in 2004 to 18.9% in 2018. Wind and solar energy sources potential are enormous. The raw potential for wind energy in Europe, especially for offshore installations, is immense, and solar radiation quantity that reaches the earth is a lot higher than the annual global energy consumption. The exploitation of this energy source depends on the geographical location of the region, land availability, and weather conditions. Several EU states are well located for the exploitation of solar energy.

4.2 Geographical Background

The European continent is located in the northern hemisphere and is recognised as the second smallest continent of all, but the third most populous after Asia and Africa. It is actually part of a huge land mass that some authors call Eurasia, but due to its particular historical and cultural characteristics, it has been separated from Asia.

Europe is surrounded by the North Polar Cap and the North Cape in the north, the Atlantic Ocean in the west, the Mediterranean, the Black Sea and Caucasus Mountains in the south, and the Ural Mountains and the Ural river in the east.

The population in Europe lies between 739 and 743, that is, about 11% of the world's population. In total, the continent covers about 6.8–7% of the land area with approximately 10,359,358 km^2.

The European topography consists of a large central plain surrounded by mountainous regions. Most of the territory (66%) does not exceed 200 m of altitude; however, the influence of the mountains is very significant. The highest mountains of the European continent are Elbrus (5633 m), Mount Blanc (4807 m), and Mulhacen

(3478 m). The mixture of plains and mountainous areas has given the territory a very valuable ecological diversity, making it possible the exploitation of several resources.

The constant presence of the sea is the most outstanding feature of the European relief. Europe has over 43,000 km of coastline and every point of the continent is near enough from the sea to receive its influence and allows the entry of the marine air masses much less contrasted than the continental ones. Europe has four seasons clearly differentiated and the climate is sweet and pleasant [1].

The four main landforms that conform to Europe are the Alpine region, Central Uplands, Northern Lowlands, and Western Highlands. Each of them characterises a different physical part of the continent. The wide diversity of physical environment provides Europe with a rich biodiversity, natural resources, and raw materials for human activities [1, 2].

The current hydrographic network is outlined after the last ice age, in which the current fjords, lakes, and large basins are formed. European rivers are moderately long, without large avenues, and generally regular. The Danube River is the longest, which is 2860 km long and flows into the Black Sea. The second longest river is the Rhine, which is 1326 km long and flows into the Atlantic. The Elbe, the Vistula, the Loire, and the Tagus are also longer than 1000 km; all of them flow into the Atlantic, except the Vistula which flows into the Baltic.

The European rivers are intensively used not only as a way of connecting different places but also to provide water for irrigation, cities, and the production of electricity. Most of the rivers are regulated by reservoirs [3, 4].

Due to its latitudinal situation between 35° and 60° lat. N, Europe is immersed in the zone of temperate climates. They are, in general, pleasant climates and adequate to today's agriculture. It is affected by the maritime polar and continental polar air masses, but also the maritime tropical air masses, and only occasionally tropical continental. The constant presence of maritime air masses makes Europe a stable and pleasant climate [5].

One of the factors that participate most in this moderating action is the presence of the Atlantic Ocean, whose influence goes deep into the continent thanks to its shortened coastal profile and the open and wide northern plains. The warm current of the North Atlantic Drift, a northern extension of the American Gulf-Stream, contributes greatly to this Atlantic regulatory action, which softens the temperatures of the northern European coast. Added to this is the regulatory role that the Mediterranean Sea plays on the coasts of southern Europe. Europe is immersed within the zonal circulation of the west flow, whose penetration into the continent is favoured by the west–east orientation of the large relief lines that do not offer serious obstacles. However, the existence of very jagged coasts, the arrangement, and structure of the landforms, the diversity of exposures in the face of moisture-laden winds, are factors that imply an enormous variety and contrast within the general moderation that characterises the European climate.

The four most important climates in Europe are the Mediterranean climate, the maritime climate on the west coast, the dry climate of the middle latitudes, and the humid continental climate. However, in the northern regions, it is also possible to find the climate of the boreal forest and the tundra climate [5].

4.3 Historical Background

On these ecological environments, favourable in most cases, a great diversity of cultural landscapes of anthropic origin has emerged, a faithful reflection of the rich and complex history of human occupation of European territory. Today, the "old continent" is a mosaic of people and cultures, the result of man's adaption to the natural environment that has been transformed by the millennial action of each people in their regional environment. The geographic diversity of the continent shapes a multicultural and diverse population in the continent, which forces unites and divides it.

The European continent is well known as the birthplace of Western Civilisation. Its rich cultural heritage and long history of human progress help in the solidification of the European Community. Its cultural wealth, with a very large number of nations and nationalities in such a small area, is considered one of the continents greatest possessions and it continues to flourish nowadays. Individual cultural identities and products are strongly supported in Europe.

There are 23 official languages in the European Union and over 60 indigenous languages across the continent. Europe strongly supports multilingual education and has adopted a multilingual policy in order to reinforce the relationship between the diversity of people and cultures. The aim of this policy is to reinforce educational, social, professional, and economic relationships in Europe and this way improves the competitiveness in global markets [6].

The recent history and political geography of Europe have been dramatically affected by World War I (1914–1918) and World War II (1939–1945) [7, 8].

Around 43 million Europeans died during World War II and left more than 40 million refugees; it also helped in the way to independence of numerous European colonies in the world and destroyed the urban infrastructure of numerous European cities. After World War II, Western Europe's leadership in global politics was weakened, and therefore the United States started to lead the Western world. [7]. In the meantime, the Soviet Union, with its capital Moscow in Eastern Europe, led the Eastern Block. The confrontation between the United States with a free-market economy and the Soviet Union, with a communist economy, was known as the Cold War. [7, 9, 10]

During the decade of the 60s Europe experienced economic growth. The European Union countries stopped charging custom duties when trading between them and the surplus agricultural produce was also soon established [11, 12, 13].

In 1973, Ireland, Denmark, and the United Kingdom joined the European Union, enlarging for the first time the community. The same year, there was an energy crisis and also economic problems caused by the brief yet brutal war between Arabs and Israel. Large amount of money starts to be transferred by the European Union regional policies to create infrastructure and jobs in poorer areas. The awareness related to pollution also increases intensifying the fights against it [11].

In the decade of the 80s, first Greece and then Spain and Portugal, joined the European Union, with 12 member states. In 1986 the Single European Act, which provided the basis for a 6 year program with the goal of having free flow of trade across the EU borders and therefore "Single Market", was signed [11, 14].

The economic and political decline of the Soviet Union brought the end of the Iron Curtain, symbolized by the falling of the Berlin Wall in 1989. [15]. There were several anti-communist revolutions and protest in this period [5]. This lead to the reunification of Germany after 28 years [11, 16, 17].

With the downfall of communism in Eastern Europe and the end of the Cold War, European citizens became closer neighbours. The "Single Market" was finished in 1993 and it established the four freedoms: free movement of people, goods, services, and money [11]. Europe started to be known as the European Union (EU), an economic and political organisation created after the Maastricht Treaty in 1993 with the aim of creating a unified structure for social, environmental, military, and economic policies for its member states [18, 19].The concern about environmental protection also increased and Europeans started to act together in matters of security and defence [20, 21].

In 1995 Austria, Finland and Sweden joined the European Union. Thanks to the Schengen agreement people gradually started to travel without using their passports checked at the borders and millions of young people started to study in foreign countries with the EU support [22, 23].

In the 2000–2010 decade the euro became the new currency for several Europeans. The euro is currently the second most popular currency after the American dollar. It is daily used by more than 320 million people [24, 25, 26].

In 2004 10 new countries joined the EU and in 2007 Bulgaria and Romania also became part of the union. The political differences between the eastern and western parts of Europe have finally been declared healed. The treaty of Lisbon came into force in 2019 providing the member states modern institutions and more efficient ways to work [11, 27, 28, 29].

The last decade (2010–2019) is marked by the global financial crisis, which started around 2008. This economic crisis was defined by debt and high unemployment. The EU members established the "Banking Union" to ensure more reliable and safer banks and helped many member states to confront their financial difficulties [11], and created a $957 billion "rescue package" to help countries with unsustainable debt rates. However, the "rescue package" led to tensions between economically competitive and without debt countries which were helping the most damaged ones [30].

In 2012 the European Union received the Nobel Peace Prize. Croatia joined the union in 2012, becoming the 28th country. The importance and concerns related to climate change and reduction of harmful emissions are high on the agenda. After the global financial crisis, as well as concerns about the relationship between immigrants and religious extremism and terrorism, has led to a more guarded attitude to immigrants [31, 32].

The United Kingdom left the European Union on 31 January 2020, after 47 years of EU membership. There are currently 27 member states in the EU. The relationship between the United Kingdom and the European Union is regulated through four treaties which define the Withdrawal, Trade and Cooperation, Information Security and Cooperation for the safe and peaceful use of Nuclear Energy Agreements between them [33, 34, 35].

The fast development of the European Union, though, has generated a number of administrative and political strains. Critics think the process of achieving EU membership is too hard for Europe's developing economies. Severe EU regulations make it very difficult for developing countries to compete with their more developed neighbours. [36]

Textbox 4.1 History background in the EU. *Source* own elaboration and adaptation from [37]

1950–1959	The start of cooperation in Europe
1960–1969	Economic growth
1970–1979	The first expansion of the European Union
1980–1989	The fall of the Berlin Wall
1990–1999	Withdrawal of frontiers
2000–2009	Further enlargement of the EU
2010–2019	A challenging decade

4.4 Economics

Europe strategically favourable location and its adequate climates, together with its easy access to several rivers and seas help in the progress of the Industrial Revolution, which was a key in the increase of very technical and urban societies. Europe became one of the fundamental economic and financial centres of the global economy. It is important to mention that the high human population densities together with the urbanisation of Europe, has led to a strain on the natural environment causing important deforestation and loss of natural habitat, which has generated a loss in biodiversity.

In terms of economy, the total value of all services and goods (GDP) of the EU is bigger than the US. The GDP of the European Union in the year 2017 was 15.3 billion euros.

Figure 4.1 describes the evolution of the per capita GDP during the last years. The value of the GDP per capita suffered a decrease during the financial crisis in 2009 and again in 2020 due to the covid-19 pandemic. The GPD varies notoriously from one EU member to another as can be seen in Table 4.1.

At least 64% of all the trades of the EU are carried out between EU countries. As mentioned before, just 6.9% of the world's population live in the EU, however, its exports and imports with the rest of the world represent 15.6% of the world's trade. The United States, China, and the EU are the three most important economies in international trade, in 2016 for example the EU countries had the second largest

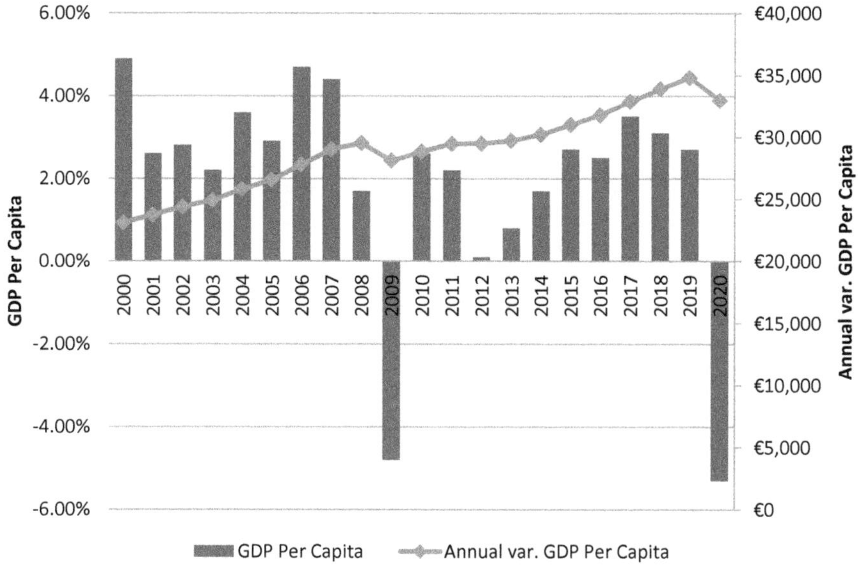

Fig. 4.1 Annual GDP per capita and annual var. GDP per capita. *Source* Own elaboration and adaptation from Expansion/Datosmacro.com

share of exports and imports in the world merchandise. In 2014 China exceeded the exports of the EU for the first time in its history. Regarding the imports, the United States had a larger share of world imports than the EU countries and China, the share is 17.6%, 14.8%, and 12.4% respectively [36, 37].

Europe's population is increasing due to natural growth (the annual birth rate is higher than the death rate) and immigration (there are more people coming to the EU than people leaving) (Fig. 4.2).

As life expectancy increases and the number of birth decreases, the population of Europe is ageing (see Fig. 4.3) [38, 39, 40, 41].

The per capita electric power consumption relates to the internal energy consumption and population. It provides not only a good perspective of how prosperous a country is but also delivers information about the improvements in energy efficiency and reduction of CO_2 emissions (Fig. 4.4).

Europe is one of the most urbanised continents in the world. Currently, over two-thirds of the European population lives in urban areas. Cities are places where connectivity, creativity, and innovation have place and are the service centres for the surroundings. Therefore, cities have a key role to activate the economy. Apart from this, since cities are places densely populated, they offer enormous potential to save energy and move towards a carbon–neutral economy. However, they are also places where problems such as unemployment, segregation, and poverty are concentrated [42, 43, 44].

The common vision of the European city for the future regarding the purposes, objectives, and values include:

Table 4.1 GPD per capita in EU members [39]

EU member	GDP per capita
Bulgary	51
Croatia	64
Romania	66
Greece	67
Laltvia	69
Hungary	71
Poland	71
Slovakia	71
Portugal	78
Estonia	82
Lithuania	82
Slovenia	87
Cyprus	91
Spain	91
Czech Republic	92
Italy	97
Malta	99
France	104
Finland	112
Belgium	118
Sweden	120
Germany	123
Austria	128
Denmark	129
Netherlands	130
Ireland	191
Luxembourgh	261

- A place of advanced social progress, with socially balanced accommodation, and with social, health, and "education for all" services
- A place for cultural dialogue, diversity, and democracy
- A place of green, ecological or environmental regeneration
- A place of attraction and core of economic growth.

Demographic change involves some challenges that differ between cities, such as the ageing of the population [41], the depopulation of cities [45, 46], or the intense processes of suburbanization [47].

Europe is no longer in a situation of continuous economic growth and many cities are facing a serious threat of economic stagnation or decline, especially those of

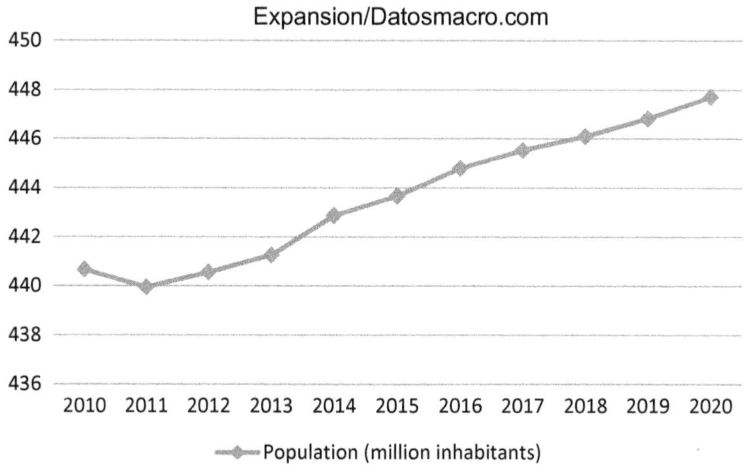

Fig. 4.2 Evolution of the population. *Source* Own elaboration and adaptation from Statista.com

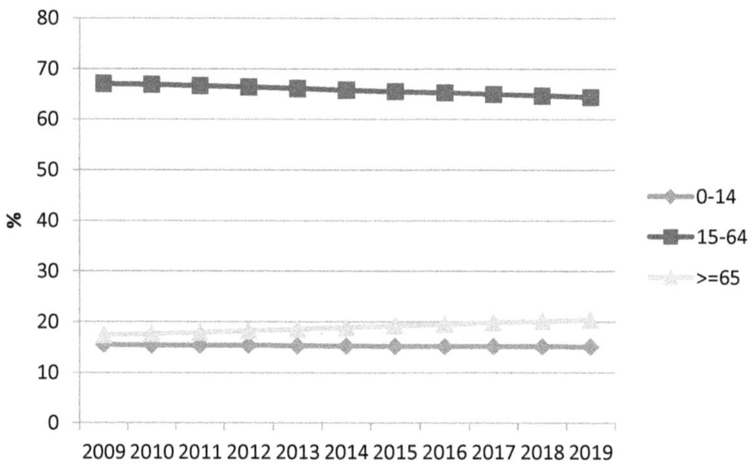

Fig. 4.3 Age distribution in the EU. *Source* Own elaboration and adaptation from Statista.com

Central and Eastern Europe that are not capital cities and also the former industrial cities of Western Europe.

The economies in their current form are not capable of providing jobs for everyone. Increasingly weak links between economic growth, employment, and social progress have generated large numbers of the population out of the labour market or into low-skilled and low-paying jobs.

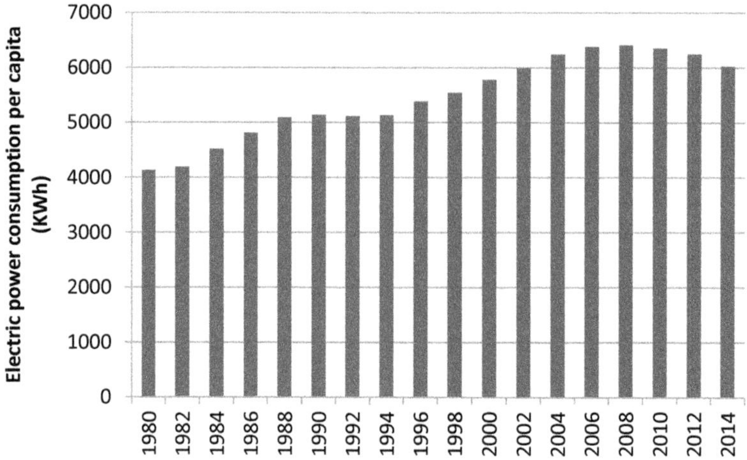

Fig. 4.4 EU's per capita electric power consumption. *Source* Own elaboration and adaptation from Worldbank.org

Increase in income disparities and impoverishment of the poor. In some neighbourhoods, local populations suffer inequalities in terms of substandard housing, poor quality education, unemployment, and difficulty or inability to access certain services (health, transportation, ICT).

Increased social polarization and segregation: The recent financial crisis has further augmented the effects of market developments and the gradual decline of the welfare state in most European countries. Social and temporal segregation are growing problems, even in our wealthiest cities [48].

Spatial segregation processes (as a consequence of social polarization) make it increasingly difficult for low-income or marginalized groups to find decent housing at affordable prices.

A growing number of "social outcasts" can lead in many cities to the development of closed subcultures with fundamentally hostile attitudes towards the majority society.

Urban expansion and the spread of settlements with low population density are one of the greatest threats to sustainable territorial development since public services are more expensive and more difficult to offer, natural resources are overexploited, public transport networks are insufficient and the dependence on cars and road congestion in and out of cities are very strong.

Urban ecosystems are under pressure. Urban sprawl and soil sealing threaten biodiversity and increase the risk of floods and water scarcity [49].

4.5 Energy Resources

Regarding natural resources, the beneficiary countries on the European continent are Ukraine, Russia, and Kosovo, as they have abundant natural resources. Natural gas and petroleum can be found in Eastern Europe. This area is also rich in carbon and iron mineral, principally in Poland and Russia, copper, silver, zinc, gold, bauxite, and nickel.

Northern Europe is rich in metals, such as bauxite (to produce aluminium), cupper, or iron mineral. There also can be found petroleum reserves, principally in Denmark and Scandinavia. Gas natural is also available in Denmark.

In Western Europe the following resources can be found: carbon, principally in Spain, France, and the United Kingdom, zinc and copper in Spain, France, and Portugal, lead in Spain and France, bauxite in France, Uranium in France and Portugal, petroleum and natural gas in the United Kingdom and Ireland, and gold in the United Kingdom and Portugal.

In Southern Europe carbon, zinc, petroleum, bauxite, iron mineral, lead, and mercury are the most common resources, especially in Italy, Greece, Croatia, and Bosnia [50].

One of Europe's biggest problems is its energy dependence. Western Europe is totally lacking in fossil fuels. It was rich in deposits of coal, but with the arrival of other fossil fuels such as oil and natural gas and with the advent of nuclear energy, the coal reserves became obsolete. Today it is trying to remedy this situation by diversifying its sources of supply [51].

The 58% of the energy consumption in the EU comes from imported energy. In 2018, petroleum products consisted of around two-thirds of the imported energy, followed by natural gas (24%) and solid fossil fuels (8%).

The EU is mostly dependant on Russia for the import of fossil fuels such as crude oil, natural gas, and solid fuels and it also depends on Norway for imports of crude oil and natural gas. The dependency rate on imports has increased since the year 2000 when it was 56% to the current rate of 58% [49].

The figure below shows the petroleum reserves in Europe. It clearly shows that the reserves are very low and the dependence on other countries very high (million barrels) [52].

On the contrary, the potential of renewable energy sources such as solar and wind energy is enormous.

Leaving aside some of the environmental, social, and economic considerations, the raw potential for wind energy in Europe is enormous. Projections of wind energy claim that the wind turbine technology potential could be equal to almost 20 times the energy demand in 2020 [53].

The highest potential is found in offshores, especially in the Atlantic Ocean. This area is affected by two semi-permanent large-scale features, the Azores high and the Icelandic low. This generally causes a big pressure gradient which is related to frontal systems moving to the east across the ocean.

Concerning the onshore wind energy potential, the maximum, even though much lower than the offshore, can be found over Denmark, Iceland, and the British Isles [54].

The theoretical potential of solar energy is enormous. The solar radiation quantity that reaches the earth is a lot higher than the annual global energy consumption. The large-scale exploitation of this energy source depends on the location of the area, land availability, and weather conditions.

With solar energy, it is possible to generate electricity, heat, cold, steam, light, ventilation, and even hydrogen. Although its high versatility, efficiency, and low cost, high efficient end-use technologies and effective energy storage need to be available.

Photovoltaic systems are one of these technologies. They convert solar light directly into electricity using a number of commonly crystalline silicon cells connected in series. The efficiency of these solar modules is 12–15% [54, 55].

Solar thermal systems can generate electricity from the high temperature they produce. Parabolic dishes and parabolic trough systems and solar power towers are the most common technologies used to generate electricity by solar heat. They use tracking mirrors reflecting solar radiation onto a power convertor unit [55].

4.6 Energy Production

The European Union depends both on its own production and imports of energy to stock up the current energy demand. In 2018, the EU imported 58% of the energy while 42% was self-produced. The figure below (see Fig. 4.5) shows the energy mix in the EU for the year 2018. [49].

It is worth mentioning that around 72% of the energy is produced from fossil fuels. This means that the dependency on these fuels is still very high. The energy mix of different member states in the EU varies considerably. Cyprus, Malta, and Luxembourg, for example, are highly dependent on petroleum products, being 91%, 87%, and 65% of the energy source respectively. The third part of the energy mix depends on natural gas in Italy 37%) the Netherlands (35%). Estonia (72%) and Poland (47%) depend mainly on solid fossil fuels such as coal and nuclear energy representing 42% in France and 32% in Sweden. Latvia and Sweden are pioneers in the use of renewable energy sources, being 39% of the mix in both countries. [49]

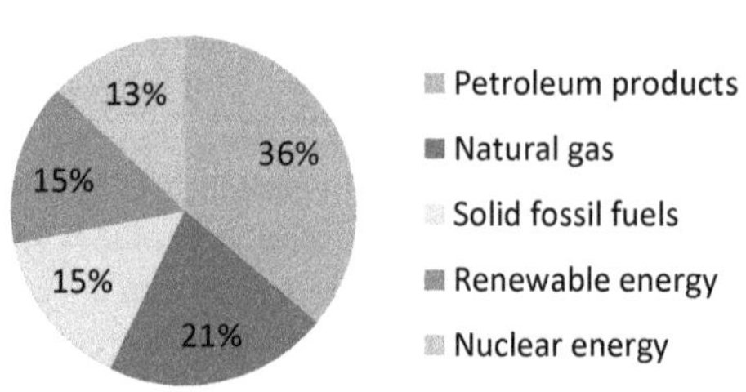

Fig. 4.5 Energy mix in the EU in 2018. *Source* Own elaboration and adaptation from ec.europa.eu

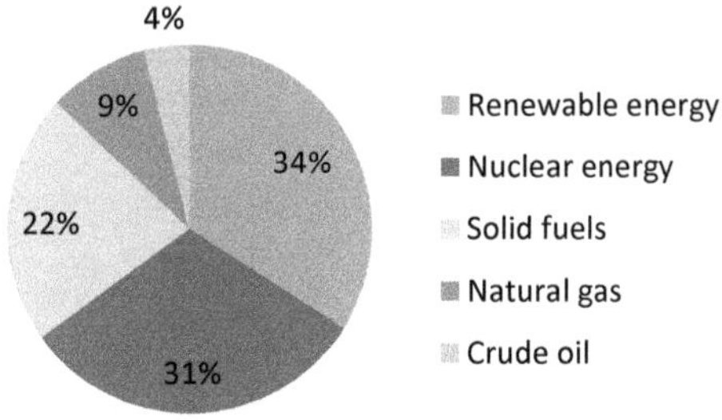

Fig. 4.6 Self-produced energy mix in 2018. *Source* Own elaboration and adaptation from ec.europa.eu

4.6.1 Energy Production in the EU

The European Union depends both on its own production and imports of energy to stock up the current energy demand. In 2018, the EU imported 58% of the energy while 42% was self-produced. The figure (see Fig. 4.6) shows the produced energy mix for the year 2018.

However, energy production varies considerably between member states. France produces 78% of its national energy from nuclear power, Belgium 65%, and Slovakia 63%. In several member states, nuclear energy is the main source of energy production with over 90% in Latvia, Malta, Cyprus, Portugal, and Lithuania. Solid fuel production is the highest in Poland, Estonia, Greece, and Czech Republic with 78%, 74% 57%, and 55% of energy production respectively. Natural gas is the main source of energy production in the Netherlands (76%) and Denmark produces 41% of its energy from crude oil [56, 57].

4.6.2 Renewables

The renewable power consumption augmented from 9.6% in 2004 to 18.9% in 2018. Sweden has the highest share of renewables among the EU member states by far with the 54.6% of the energy consumption. It is closely followed by Finland (41.2%), Latvia (40.3%), Denmark (36.1%), and Austria (33.4%). The lowest share of renewable energy sources in the final consumption was found in the Netherlands (7.4%), Malta (8%), Luxembourg (9.1%), and Belgium (9.4%) [58].

For the first time, in the first half of 2020, the EU generated more electricity from renewable sources than from fossil fuels. Wind, solar, hydro, and bioenergy produced 40% of the electricity among the member states, while fossil fuels generated 34% [59, 60].

The energy generation was not the same in the different member states; Denmark was on top producing 64% of its energy from renewable sources, followed by Ireland with 49% of the generation and Germany with 42% [60].

The expansion in the use of renewable energy sources in the EU, especially wind, solar and solid biofuels (including renewable wastes) is visible in the increase in

Table 4.2 Electricity produced from renewable sources

Country	%	Country	%
Austria	73	Greece	26
Sweden	66	Bulgaria	22
Denmark	62	Slovakia	21
Latvia	53	France	21
Portugal	52	Estonia	20
Croatia	48	Belgium	19
Romania	42	Lithuania	18
Germany	38	Netherlands	15
Finland	37	Czchia	14
Spain	35	Poland	13
Italy	34	Cyprus	9
Ireland	33	Luxembourg	9
Slovenia	32	Hungary	8
United Kingdom	31	Malta	8

the amount of electricity generated from green energy between 2008 and 2018. The electricity generated using solar and wind energies were 15.5 and 2.9 times higher in 2018 than it was in 2008. The electricity generated using solar energy increased dramatically from 7.4 TWh in 2008 to 115.0 TWh in 2018 [61].

The quantity of renewable energy sources each EU member state is using to produce electricity varies significantly. In Austria, Sweden, and Denmark for example the 73.1%, 66.2%, and 62.4% of the electricity respectively are generated using renewable energy sources. On the contrary, in Cyprus, Hungary, and Malta the share of electricity produced from renewable sources is 9.4%, 9.1%, and 8.3% respectively, less than 10% [62] (see Table 4.2).

Regarding the energy sources used for cooling and heating in the European Union, in the year 2019, 20.5% of the energy was generated using renewable sources. The figure below shows the increase in the renewables share during the last 10 years (see Fig. 4.7). Growths in services, industrial sectors, and households have helped to this increase [63].

The amount of renewable energy sources used in the different state members is shown in Table 4.3.

Figure 4.8 shows the increase in the average use of renewable energy sources for transport in the European Union in the last years.

The use of these fuels varies considerably among different state members. In Sweden for example the 29.7% of the energy used in transport comes from renewable sources, while in Croatia only 2.6% of the energy for transportation is produced with green energy sources. The table below (see Table 4.4) shows the renewable energy share for the different state members. [64].

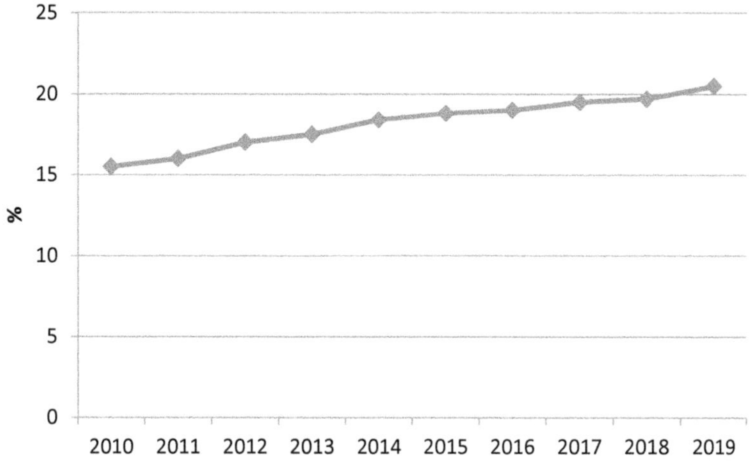

Fig. 4.7 Renewable energy share in the EU

Table 4.3 Renewable energy share in EU members

Country	%	Country	%
Sweden	66.1	Malta	25.7
Latvia	57.7	Czechia	22.6
Finland	57.5	France	22.5
Estonia	52.3	Slovakia	19.7
Denmark	48	Italy	19.6
Lithuania	47.3	Spain	18.8
Portugal	41.,6	Hungary	18.1
Croatia	36.8	Poland	15.9
Bulgaria	35.5	Germany	14.5
Cyprus	35.1	Luxembourg	8.7
Austria	33.8	Belgium	8.3
Slovenia	32.2	United Kingdom	7.8
Greece	30.2	Netherlands	7.1
Romania	25.7	Ireland	6.3

In some European Union members such as the Netherlands, Sweden, Finland, Malta, Luxembourg, and Ireland for example, the acceptance of renewable energy as a transport fuel was very high.

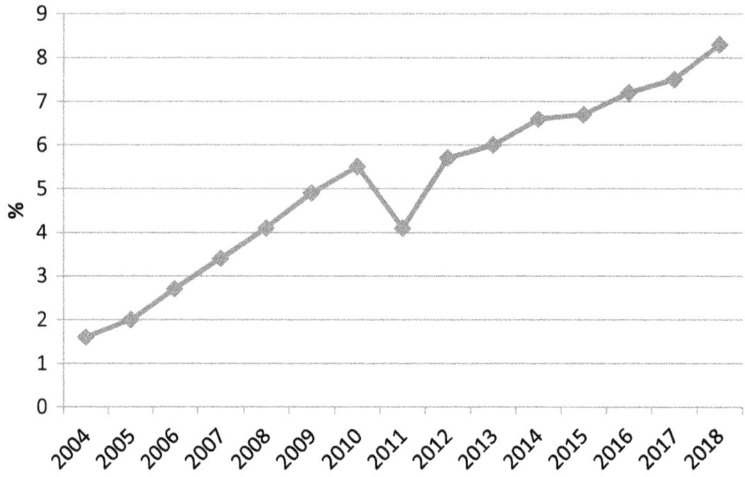

Fig. 4.8 Renewable energy share in transport

Table 4.4 Renewable energy share for transport in EU members

country	%	Country	%
Sweden	29.7	Denmark	6.9
Finland	17.7	Belgium	6.7
Austria	9.9	Luxembourg	6.6
Netherlands	9.6	Czechia	6.6
Portugal	9	United Kingdom	6.5
France	9	Romania	6.3
Malta	8	Poland	5.6
Bulgaria	8	Slovenia	5.5
Germany	7.9	Latvia	4.7
Hungary	7.8	Lithuania	4.3
Italy	7.7	Greece	4.1
Ireland	7.2	Estonia	3.3
Slovakia	7	Cyprus	2.7
Spain	7	Croatia	2.6

4.6.3 Environmental Impact

Global warming and climate change issues are probably the greatest threats to this planet and most controversial topics of these present times. After years of intensive research, the scientific community agrees that the excessive use of fossil fuels is the most relevant cause of climate change [65]. The emissions related to the burning of these fuels are concentrating in the atmosphere and leading to the warming of the

Earth. Apart from the hazardous effects of the utilisation of fossil fuels, it is important to note that their reserves are finite, and the enormous consumption of oil, gas, and coal in the last 200 years has led to their depletion [55].

In the European Union, the greenhouse gas (GHG) emissions decreased gradually between 2010 and 2014, however, after that the trend changed and started to increase again [65]. It is worth mentioning that in 2017 the GHG emissions in the EU were 19% lower than the emissions in 1990, representing a reduction of 935 million tons of CO_2 equivalents [66] (Fig. 4.9).

In 2017, the energy producing industries had the highest amount of GHG emitted to the atmosphere followed by fuel combustion by users and transport with the 29%, 25.5%, and 23.8% respectively. Table 4.5 shows the GHG emissions by source in 2017 [66].

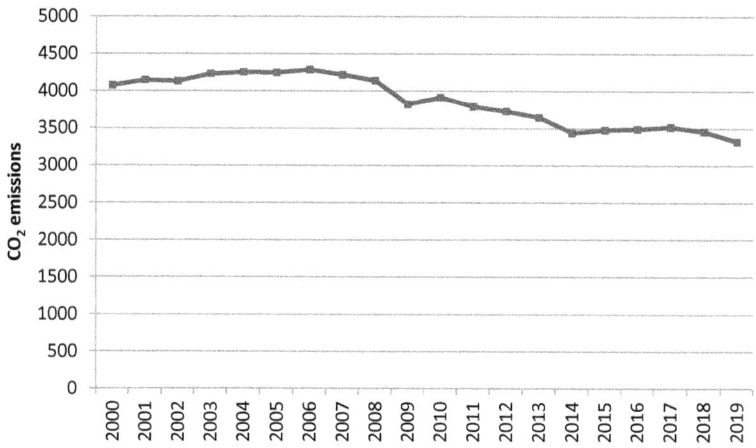

Fig. 4.9 CO_2 emissions (million tons of CO_2 equivalent)

Table 4.5 GHG emissions by source in 2017

Greenhouse gas emission by source 2017 (%)	
Energy industries	29.00
Fuel combustion by energy users	25.50
Transport	23.80
Agriculture	10.00
Industrial processes and product use	8.70
Waste	3.00

4.7 Energy Policy

The challenges that the EU is facing in the field of energy include its increasing dependence on imports, limited diversification, high and unstable prices of energy, rising global power demand, risks of security affecting producer and transit regions, growing threats arising from climate change, slow development in energy efficiency or the challenges caused by increasing the share of green energy sources, and the need to increase transparency and better integration and interconnection of energy markets. The core of European energy policy is made up of a series of procedures with the objective of getting an integrated energy market, security of energy supply, and sustainability of the energy sector [67].

The EU, a pioneer in renewable energy, has made considerable efforts in its legislation to better integrate renewable sources into European energy systems. In its quest to be a world leader in this field, the EU has established a clear route for others to follow. When the Renewable Energy Directive (2009/28 / EC) set national objectives for member states, the measure was considered “unprecedented”. Today there are 173 countries in the world that have similar goals.

The current framework for action is based on the general integrated policy on climate and energy adopted by the European Council on October 24, 2014, and revised in December 2018, which establishes these objectives for 2030 [68]:

- a decrease of at least 40% of GHG emissions compared to 1990 levels;
- a 32% increase in renewable energies sources for the energy consumption;
- an improvement in energy efficiency of 32.5%.

Even more ambitiously, the European Green Deal proposes a series of initiatives in all areas to achieve a climate neutral EU by 2050. Renewable energy sources and energy efficiency are the key factors in the achievement of such ambitious goal. The European Commission proposes new measures in order to incorporate technological advances in all the areas related to the energy system. This initiative, also known as “smart sectoral integration” will help create tomorrow’s European energy system [69].

The Commission has elaborated a new strategy to improve and boost offshore wind energy. This strategy will address the challenges and opportunities, the management of the maritime space, the impact on the energy markets and networks, and the industrial policy matters of this energy source [70].

Textbox 4.2 European legislation in renewable energy. *Source* own elaboration and adaptation from ec.europa.eu

1991	First feed-in tariff for renewables introduced in Germany
1997	New target of the use of 12% of renewable energy sources by 2010 in the EU
2000	First large scale offshores wind farm built in Denmark
2001	Directive on electricity generation from renewable energy systems through national indicative targets

(continued)

(continued)

1991	First feed-in tariff for renewables introduced in Germany
2003	Directive on biofuels and renewable fuels for transport through national indicative targets
2008	The largest solar PV power plant (60 MW) in the world built in Spain. Generates enough to power 40,000 homes/year
2009	Renewable energy directive: EU target of 20% renewable sources by 2020 and national binding targets
2014	Onshore wind energy becomes cheaper than coal, gas and nuclear energy
2018	Revised Renewable Energy Directive: EU target of 32% renewable sources by 2030
2019	EU power production from solar and wind energy exceed coal for the first time

4.7.1 European Green Deal

The European Green Deal (COM (2019) 640 final) is the ambitious EU climate policy that has the objective to become Europe the world's first climate-neutral continent by 2050. The measures taken related to this deal have the goal to enable European citizens and businesses to benefit from sustainable green transition [58]. For this purpose it is essential to transform the European energy system, providing secure, sustainable, competitive, and affordable energy to the EU citizens and businesses [71].

The usage of renewable energy decreases greenhouse gas emissions, diversifies energy suppliers, and decreases dependency on fossil fuel markets (especially oil and gas). Besides, increasing renewable energy sources also helps in the development of employment in the creation of jobs related to green technologies [58].

The most important policy instrument is the Energy Union, but it is not the only one. In the following section, different strategies are explained [71].

Energy Union

To guarantee a reliable energy supply and preserve affordable prices, the European Union wants to create a more integrated and competitive European energy market (Energy Union). The EU supports energy from renewable sources and the efficient use of energy, all to reduce greenhouse gas emissions [72].

The objective of the Energy Union Strategy (COM/2015/080), published on 25 February 2015, is to build an energy union to provide EU consumers sustainable, competitive, secure and affordable energy. Since its launch, numerous packages with measures and progress reports have been published to make sure the energy union strategy has been achieved [73].

The Energy Union Strategy established five main interconnected and mutually reinforcing objectives, in order to increase energy security, sustainability, and competitiveness.

- Energy security, solidarity, and trust:
  Diversify EU's energy sources and increase the efficiency of the energy produced in the EU [9]. The EU needs to be more independent on imported energy by making more efficient use of its own energy and diversifying sources and supplies at the same time [72].
- A fully-integrated internal energy market:
  Remove technical and regulatory barriers among different countries in order to enable free energy flow in the EU members. This way, energy providers will be able to freely compete and provide the best energy prices [9]. The EU wants fewer technical and legal obstacles so that energy can cross national borders and suppliers can compete throughout the EU. The Trans-European Networks Strategy (TEN-E) aims to expand and update European infrastructures and create networks across national borders [72].
- Energy efficiency contributing to moderation of demand:
  Reduction in energy usage will reduce pollution and reserve domestic energy sources. This will also help in the reduction of the dependence of the EU's energy imports [71]. EU standards on construction, industry, consumer products, and transport are helping the EU achieve its energy efficiency goals and move towards a low-carbon society [72].
- Decarbonising the economy:
  Encouraging a global climate change deal and reinforcing the investments of private sites for new infrastructures and technologies [71]. EU regulations want to maintain fair markets for fossil fuels and protect the environment, including when applying new technologies such as shale gas extraction. The EU coordinates efforts to achieve national objectives that are in line with the Renewable Energy Directive. It also promotes the use of alternative energy in transportation [72].
- Research, innovation, and competitiveness:
  Reinforcing the research on low-carbon technologies and supporting financially projects in partnership with the private sector [58]. The EU supports the spread of low-carbon technologies such as solar, wind, carbon capture and storage, and energy storage technologies [72]. The Horizon 2020 (H2020) program, which covered the period 2014–2020, and Horizon Europe, which covers the period 2021–2027 are the Union's main instruments for promoting energy research. Funds of up to € 95.5 billion have been earmarked for clean, safe, and efficient energy and sustainable development.

Clean energy for all Europeans package

The Clean energy for all Europeans package was created as a consequence of the EU's Paris Agreement commitments for reducing greenhouse gas emissions, reducing fossil fuel use, and increasing the use of cleaner energies. This package helped considerably in the implementation of the Energy Union Strategy. The Clean Energy for all Europeans package consists of eight legislative acts and with the help of the political agreement between the Council and the European Parliament and the entry into force of the different EU rules, the new directives will transpose into national laws in the coming years [74].

These new guidelines will bring benefits from consumer, environmental and economic perspectives. This change at the EU level also emphasizes EU leadership in fighting global warming and offers an important influence to EU's long-term strategy of achieving carbon neutrality by 2050 [73].

Clean energy annual events

EU Energy Day

Meeting key clean energy groups to strengthen and promote cooperation on sustainable, secure, and smart energy solutions.

EU Sustainable Energy Week

EUSEW is the greatest event in Europe on renewables and energy efficiency, bringing together policymakers, industries, and citizens [73].

National energy and climate plans (NECPs)

In order to achieve the EU's climate and energy objectives for 2030, EU Member States need to fulfil the EU's energy and climate objectives for 2030. Therefore, EU Member States need to create a 10-year integrated national energy and climate plan (NECP) for the period from 2021 to 2030 [73].

2050 Long-term strategies

In order to meet the economic transformation needed and wider sustainable development objectives, it is essential to have stable long-term strategies. This also helps to achieve the long-term objective established in the Paris Agreement to keep the increase of the mean temperature below 2 °C compared to pre-industrial levels and to limit the temperature increase to 1.5 °C [73].

4.8 Wind Turbine Installations

The European Union has been a pioneer in the establishment of renewable energy systems becoming the global leader. Wind power has been one of the most important energy sources in this success and it will be indispensable to reach the EU's renewable energy goals and achieve carbon neutrality by 2050 [75]. Wind is clean, free, and abundant and it is a really competitive source of electricity [65]. The electric current is generated when wind turbines capture the kinetic energy of airflows. Wind farms are constituted by several wind turbines, usually covering many square kilometres of land (onshore wind energy) or sea (offshore wind energy) [75].

Development and innovation in the wind energy sector have led not only to improvements in the design and manufacturing of turbines but also to better capacity factors, causing the drop of costs of wind power and confirming this technology as a key driver of clean energy transition [75]. Wind energy is currently the cheapest method of new power generation in Europe [76].

It is also worth mentioning that wind energy is in Europe one of the key contributors to the economy in terms of increasing growth and generating long-term sustainable jobs [75]. Wind power accounts for 72€ billion annual turnovers and around 330,000 jobs [76]. It is estimated that in 2017 around 1.45 million people were working in the renewables sector in the EU, from which the wind sector provided 356,700 full time jobs [75].

The wind energy industry has grown from a niche sector to a mainstream industry. The competitiveness increases yearly and it attracts more and more investments. Three of the five largest global wind turbine manufacturers are European companies, leading to the 40% share of all wind turbines sold globally. Overall, it can be said that wind is an essential and growing sector in the European industrial base [76].

The European Union bets on Wind Energy technology to provide the largest contribution to the EU renewable energy goals from 2020. Currently, around 70% of the global wind energy capacity is in the EU. The goal for 2030 is to reach a capacity of 350 GW, supplying around 24% of the electricity demand [65].

During 2019, Europe installed 15.4 GW of new wind energy volume, from which 13.2 GW were installed in the EU28, 27% more than in 2018, but 10% less than in 2017, which was the record. [77]. Wind power installations accounted the 44% of all new power installations in Europe [76]. Onshore installations were 11.7 GW, representing 76% of the power, while the offshore wind power installation was 3.6 GW. Around half of the offshore wind farm installations were in the UK in 2019 adding 2.4 GW to its capacity, being the 74% offshore wind energy. Spain, Sweden, and France led the installation of onshore wind energy farms, installing 2.3 GW, 1.6 GW, and 1.3 GW respectively and representing 45% of the onshore wind energy installation together. It is also worth mentioning that Denmark has the largest share of wind power in its electricity generation since it represents the 48%. This figure is followed by Ireland with 33%, Portugal with 27%, Germany with 26%, and the UK with 22% of its demand [77] (Table 4.6; Fig. 4.10).

5.1 GW was installed in Europe during the first half of 2020, 3.9 GW onshore and 1.2 GW offshore. Onshore wind turbine installations were slightly higher compared to the average of the previous three years (3.7 GW) but the offshore were lower compared to the previous three years average installations (1.5 GW). Germany had the highest number of onshore installations (587 MW) although it remained below historical levels; France had the second-highest number of installations with 494 MW. Regarding offshore wind energy technology, the UK had the highest capacity installed in this period of time (483 MW), followed by Belgium (235 MW), the Netherlands (224 MW), and Germany (213 MW). In order to fulfil the objectives set out in the European Green Deal, Europe needs to install 20 GW per year [78, 79].

COVID-19 had an important effect on electricity generation and demand during the first half of 2020. The electricity demand in most European countries decreased as much as 25% in the mid-March to mid-May period. In this context, wind energy installations generated 241 TWh of electricity covering 17% of Europe's energy demand. It is also important to mention that Europe benefited from strong wind

Table 4.6 Wind power installations (MW) [77]

Country	Onshore	Offshore	Total
UK	629	1764	2393
Spain	2319	0	2319
Germany	1078	1111	2189
Sweden	1588	0	1588
France	1336	0	1336
Greece	727	0	727
Belgium	207	370	577
Ireland	463	0	463
Italy	456	0	456
Denmark	28	374	402
Finland	243	0	243

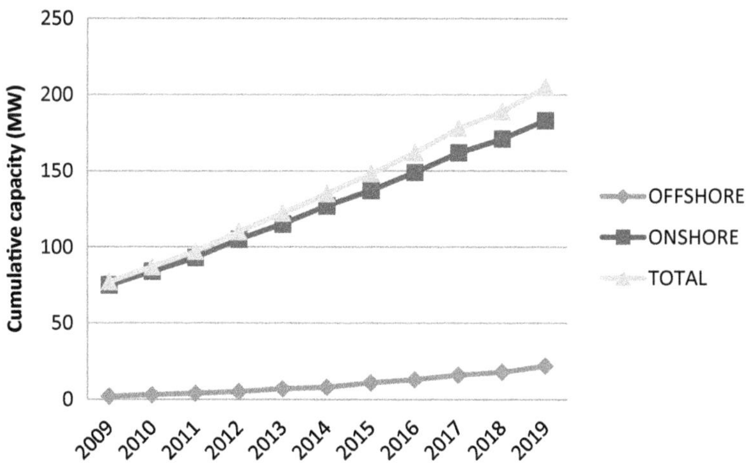

Fig. 4.10 Wind power capacity in EU (MW) [77]

generation already before COVID-19 lowered energy consumption, particularly in February, when wind covered 24% of electricity demand [80].

In terms of project finance, the economic fallout from COVID-19 increased short-term debt costs and was accompanied by debt liquidity strains in the lower-rated states in Eastern and Southern Europe. Despite the difficult economic environment, in the first half of 2020, a record amount of 14,300 million euros was raised to finance new wind farms. Offshore energy had a particularly strong six-month period with 11 billion euro of funding raised. The important projects to reach the final investment decision were the 1.4 GW Hollandse Kust Zuid (Netherlands) or the 1.1 GW Seagreen Alpha and Bravo (Scotland), as well as the French offshore wind farms Saint Brieuc and Fécamp. The financing for onshore wind power fell to € 3.3 billion from € 4.9 billion in the first half of 2019 [81, 82].

The European Council agreed on a € 750 billion recovery plan to fulfil the goals of the Green Deal. 30% of all spending will go to green investments in electrification, renewable energy, storage, hydrogen production, and other activities that will boost value chains and position Europe as a leader in green technologies and climate protection. The plans must set investment and reform priorities in line with the long-term objective of EU climate neutrality by 2050. [83].

In spite of the COBID-19 ongoing pandemic, European wind turbine orders increased 74% compared to the previous year, up to a record number of 15 GW. The onshore wind turbines went up by 13% with 8.2 GW ordered. Offshore wind turbines orders increased six fold compared to 2019 [84].

Regarding the countries that ordered wind turbine installations, the United Kingdom ordered most capacity with 4.4 GW, followed by the Netherlands with 2.4 GW and Sweden with 1.4 GW. The total number of orders was 104 from 19 different countries. Regarding the onshore wind turbines, Sweden ordered the most followed by the United Kingdom and the Netherlands. In the case of the offshore wind turbine orders, the United Kingdom was the one ordering the most followed by the Netherlands, France, and Germany [85].

4.8.1 Offshore Wind Energy

As the public acceptance gets higher, offshore wind energy signifies a growing future opportunity since the resources are stable and abundant. Research and innovation are also key factors in lowering costs and increasing the performance and reliability of offshore wind technology [65]. It can significantly contribute to growing the share of renewable energy in Europe's mix. However, there are several challenges that need to be taken into accounts, such as the competing uses of the sea, costs, and grid integration [86], as well as improvements on the potential impacts of offshore turbines on the environment and public health [65].

The circularity of wind energy technology, considering the turbine's end of life and material dependency, is also another challenge that needs to be faced [65].

The EU currently leads the world in offshore wind energy with over 90% of the wind farms. 26. The first offshore wind farm was installed in Vindeby (Denmark) in 1991, and this fact gives invaluable experience by being pioneers to the EU companies [76]. Other countries such as the USA, China, Korea, and Japan have also increased their support to increase their offshore industry [65].

In Europe, the offshore wind energy industry is focused in the North Sea, which has relatively shallow waters in the southern part and at Dogger Bank, and deeper waters to the North. The wind speeds are good across most of the area and distances to the shore are up to 200 km. and optimum wind conditions [86].

However, in order to maintain the lead in the industry, the EU is supporting integrated floating systems and substructures for deeper waters and different weather conditions. The interest in other sea basins has increased in the last years, especially in the Baltic Sea and the Atlantic. The Baltic Sea has shallower water depths, lower wind speeds ant less extreme sea states. The distances to the shore are also shorter. The Atlantic, on the other hand, has mainly high wind speeds, deeper waters, more extreme sea states, and longer distances to the shore [86]. The EU is determined to boost offshore wind development and explore its potential in Europe's seas and along its coasts but always respects the ecological limits of natural resources and the interests of other sea users [75].

This will increase disposition possibilities and expand the European position in the global market [65]. For this purpose, several European developers are working on floating offshore wind turbines, with the first pilot projects in operation. The disposition of this technology is expected to be ready by the end of this decade. The EU has become a member of the North Seas Energy Cooperation (NSEC) in order to harness full energy potential of offshore wind in the North Sea with the cooperation with other countries of the region [75].

In order to keep temperature rises below 1.5 °C, the European Commission estimates the need for between 240 and 450 GW of offshore wind energy by 2050. It is expected to supply around 30% of the electricity demand with offshore wind energy and electricity will represent at least 50% of the total energy mix by 2050 [75].

Eight Baltic countries, Germany, Sweden, Lithuania, Latvia, Poland, Denmark, Finland, and Estonia, signed a joint declaration in September 2020 with the Commission to speed up the build-out of offshore wind installation in the region.

Poland aims to build-out a 5.9 GW offshore wind farm in Polish waters by 2030. For this purpose, the Polish Government adopted its Offshore Wind Bill in order to simplify the administrative procedures [87, 88].

Estonia and Latvia will run a joint auction to build-out an offshore wind farm by 2026. The objective of this plan is to produce a hub where other similar projects in the geographic area can join in. This will also allow cost saving [89, 90].

Germany and Denmark declared that they will cooperate more in offshore wind projects. They are already working on the link of their offshore farms in the "Kriegers Flak" area in the Baltic sea with grid connections back to Denmark and Germany [90].

As shown in the previously described projects, the Baltic Sea has an incredible potential for offshore wind installation. It has been estimated that this sea could host as much as 93 GW by 2050. Apart from this providing clean and economical electricity to the Baltic countries, offshore wind projects will also help in the support of the economic recovery in Central and Eastern Europe, providing long-term growth and jobs. Countries with no access to the sea will also benefit from offshore wind too [91].

South-eastern countries such as Bulgaria and Romania are researching the potential of the Black Sea. Romania is currently working on its first Offshore Wind Law [92, 93].

All the EU members have agreed to support offshore wind projects. However, governments in Central and Eastern Europe need to invest in grid infrastructure for both on- and offshore projects. The National Recovery Plans is a good opportunity for these investments [94].

Offshore wind energy was initially developed as a bottom-fixed technology, for shallow waters such as the ones in the North Sea first and the Baltic sea then. Bottom-fix turbines are more expensive when the waters are deeper. Therefore, for deep waters, such as the Mediterranean ones, floating wind turbines are more adequate. Currently, there are 62 MW floating offshore wind turbines installed in Europe from which 30 MW are in the Hywind wind farm in Scotland and 25 MW in the WindFloat Atlantic wind farms in Portugal. The perspective for the next decade includes over 7GW floating offshore projects. Concerning offshore projects in the Mediterranean, France is the leading country, since it expects to build the first three pre-commercial floating wind farms there by 2023 [95].

The potential of wind energy in Greece is also interesting especially for floating offshore wind plants. At the moment, there are 4GW of wind energy installed in Greece, all of them onshore, and they cover 12% of the electricity consumption. The Greek government is considering developing a new legislative framework to enforce offshore wind energy. This new framework will help in the investigation of proper sites of wind installations and will ensure stable financing for new infrastructures. Greece's National Energy and Climate Plan aim to install 7GW of wind energy by 2030. The offshore wind installations would help in the energy autonomy of the islands [96].

The electricity mix in Poland still depends heavily on fossil fuels, having the 70% of the electricity generated from coal and 10% from gas or oil in 2019. Regarding renewable energy sources, wind is the biggest source with 10% of the total electricity production [87, 88].

Poland aims to transform into a low-emission economy, in order to diversify its power mix and significantly reduce CO2 emissions.

The Polish Parliament agreed to adopt Poland's first Offshore Wind Act to develop offshore wind in the Baltic Sea. This act includes regulations and rules that will apply to the development of offshore wind and it defines the volumes Poland intends to develop. Currently, Poland does not have any offshore wind installations, and its objectives are ambitious. They expect to install 3.8 GW by 2030, 10 GW by 2040, and 28 GW by 2050. This would become Poland the biggest offshores wind installation market in the Baltic Sea. They plan to have the first turbines operating by 2025 [87, 88].

This Offshore Wind Act will not only reinforce the Polish economy but also help in the post-COVID recovery. Offshore wind projects are expected to activate investments and to create tens of thousands of new job opportunities, which will build a strong maritime industry around the sector which will help to revitalise Polish shipyards and ports, and industrial clusters. Poland's big ports such as Szczecin, Gdansk, and Gdynia will transform into "renewable hubs" for storage, assembly, transportation, and maintenance of offshore wind turbines [97].

4.9 Solar PV Installations

The solar energy market is constantly growing all over the world. Technologies are advancing every day to maximise energy production in the smallest possible space, so new solar panels, solar batteries, solar regulators, etc. are constantly appearing. At the same time, the number of installations around the world is growing exponentially [98].

Solar photovoltaic installations convert the energy from the sunlight into electricity. The European Union is pioneer in the spread of solar energy technology and has taken many measures to improve market uptake and become solar energy more accessible to consumers. The Clean Energy for all Europeans package is the most important of those measures. Solar power has become one of the cheapest technologies to generate electricity with the help of a solid industrial foundation. The cost of the production of solar energy decreased 75% between 2009 and 2018. The solar market is expected to keep on growing in the following years as a cornerstone of the clean energy transition [99].

In 2000, Europe already accounted for more than 70% of the total world wind energy and 20% of photovoltaic solar installations. That same year, also in Denmark, the first large-scale wind farm, "Horns Rev", made its appearance, employing numerous technologies that would later become industry standards. By 2008, Europe had become the world's largest market for solar photovoltaic energy, with more than 70% of the market. That year, in Spain, the Olmella photovoltaic park (a 60 megawatt power plant, the largest in the world) generated enough solar energy to supply 40,000 homes a year. As the production and use of renewable energy increased in the rest of the world, Europe maintained its leadership. In July 2019, Portugal created the world's lowest cost photovoltaic solar power park … a record it holds to date [100].

The year 2019 is going to be framed by the European and especially the Spanish PV sector for many years to come. According to Solar Power Europe's "EU Market Outlook 2019–2023" presented during the event 'A Solar-Centred Energy Mix to Fight Against Climate Change' held at COP25, 2019 has become one of the best years for solar energy in the European Union.

The Old Continent installed 16.7 GW, an increase of 104% compared to the 8.2 GW installed the previous year. Moreover, this year also saw the highest solar growth since 2010, when the European PV market also increased by 104% during the first solar boom, albeit at a lower level, reaching 13.4 GW [70, 101].

Spain has become by far the largest solar market in Europe with the installation of 4.7 GW of new PV installations. According to Solar Power Europe, "the sunny country returns to the continent's top solar spot, 11 years after it last held the position". [70, 101].

Rounding out the list of the top five European solar markets are Germany (4 GW), the Netherlands (2.5 GW), France (1.1 GW), and, to the surprise of many, Poland, which almost quadrupled its installed capacities in 2019 to 784 MW [70].

In total, the EU's top five solar markets were responsible for more than three-quarters of the continent's installed capacity in 2019. And looking at the data for

Table 4.7 Installed PV capacity (MW) [102]

	2018	2019
Spain	288	4680
Germany	2947	3985
Netherlands	1500	2496
France	873	1068
Poland	203	784
Hungary	410	653
Belgium	367	605
Italy	485	598
United Kingdom	286	374
Portugal	91	367

the top 10, the share rises to 93% based on a combined capacity of 15.6 GW, which is more than double the 7.5 GW added in 2018. The report notes that "despite the relatively small contribution of the other 18 EU member states, it appears that the vast majority are now moving in the right direction, even if it occurs at a much lower level. In 2019, 26 of the 28 EU markets installed more solar capacity than in the previous year".

In total, installed PV capacity in the EU now totals 131.9 GW at the end of 2019, up 14% from 115.2 GW the previous year. Table 4.7 shows the growth in capacity for the different member states in the EU [70].

There are several reasons for Europe's new solar boom. First, the key to solar growth in EU countries but also in other countries on the continent is its competitiveness. Solar energy is cheaper than any other source of power generation today, and its attractiveness only increases as the cost reduction curve continues at a much faster pace than for any other technology.

Another important factor is the deadline for member states to meet their binding national 2020 renewable energy targets. At the same time, these same countries have already started preparing to meet the 32% renewable energy target of the Clean Energy Package by 2030. With solar energy as the most widely accepted source of power generation among EU citizens, as well as the most flexible and easy to install and often the lowest cost relative to other renewable technologies, governments are increasingly embracing solar in their climate strategies.

The array of tools and technologies that support solar is immense: from renewable auctions that show that large-scale solar can win technology-neutral bids against all other technologies; to self-consumption and storage that appeal to prosumers looking to reduce their electricity bills; to new business models that are enabled by digitalisation, such as peer-to-peer electricity supply.

According to Solar Power Europe, the average scenario for the annual development of the EU solar PV market until 2023 is one of continued growth. After demand more than doubled in 2019, they forecast 26% growth, exceeding 20 GW or 21 GW in 2020. In 2021, installations are expected to reach 21.9 GW, close to the all-time high for solar installations in the EU, and even what happened in 2011 with 22.2 GW of new installed capacity. This record is expected to be surpassed in 2022 with 24.3 GW of new projects, and again in 2023 with 26.8 GW of installed solar capacity [70].

All in all, PV is shining in the European Union and will continue to be, in 'weather forecast' parlance, bright and sunny, with very few clouds on the horizon and no rain ahead [102].

Regarding the impact of the COVID-19 on the European solar energy market, they estimate that it will only be in the short term without affecting the market over a longer period of time and with a rapid recovery. Trina Solar estimates that the pandemic will cause a 10% drop in the EU market in 2020, but this will be recovered and even exceed the 2019 numbers, in 2021 with a 47% increase in installed GW compared to the previous year.

This increase would continue until almost doubling the GW in 2024, or in the best case scenario reaching three times the current numbers. These increases are also expected to be similar for the installation of energy storage in batteries.

Techno Sun says that the downturn in the solar market in Europe has been smaller or almost non-existent in Spain, where the sector has remained active despite the pandemic driven by the recent legislative change. This will provide a more consistent basis for when the global market recovers, the speed of recovery in Spain will be even greater [98].

4.9.1 EU Solar Trends

Solar energy only represents around 5% of the total electricity share of the European Union. This means that the business opportunities are huge due to the potential of solar power in the energy generation field as well as in the power, heat, and transport energy segments [70].

Industrial solar strategy

The Green Deal package is the first step towards a strong industrial strategy proposal. The European Union's objective is not only its decarbonisation and lead by example to other economies, but there is also a chance for Europe's 4th Industrial Revolution with clean energy technologies at its core. The main challenge for EU policymakers is to recognise strategic values chains to develop three important priorities:

- Increase climate ambitions
- Make sure the transition to clean technologies will generate jobs and growth in the EU countries
- Protect Europe's energy supply

The solar energy has the potential to help in the achievement of these three strategic pillars [103].

Solar buildings

Buildings represent almost half of the final energy consumption and 36% of the European Union's GHG emissions. Therefore, decarbonising the building sector would help enormously in the achievement of carbon neutrality in the EU by 2050.

The rooftop space in almost all European buildings is unused, therefore the potential of the rooftops with cost-effective solar PV would be enormous, estimated to be around 680 TWh, which means 25% of current electricity consumption. The installation of solar panels on all new and renovated buildings in the EU would save around 7 million tonnes of CO2 emissions per year. Solar roofs would also contribute to the energy efficiency measures at building level. As much as 30% of annual final energy savings could be achieved with on-site energy generation [104].

Digital solar

The digital and energy transitions should be developed in parallel. Solar energy solutions are perfect to match with digitalisation since it is the most distributed of all energy resources [70].

Solar grid integration

The grid infrastructure makes it possible to have a stable electricity system. Solar energy technologies would be helpful and useful as more flexible resources in this regard. Regarding the distribution level, self-consumption of electricity would diminish the stress and local congestions on the grid and therefore avoid expensive network investments.

Sustainable solar

Sustainability product policy labels as for example Ecodesign, Energy Label, Ecolabel, and Green Public Procurement for PV modules, inverters, and solar systems help in the development of solar industry in the European Union [105].

Solar mobility

The transportation sector is responsible for 25% of CO2 emissions worldwide. The electrification of transport is increasingly seen as the most efficient solution to reduce these emissions. Therefore, electric vehicles and electric mobility infrastructure are becoming a growing concern for private drivers, cities, and local authorities.

Renewable financing

In order to ensure Europe's decarbonisation and industrial leadership in clean energy technologies, it is inevitable to have public and private investments in this field and an ambitious green financing strategy to control these investments and make sure that future funds are focused to clean energy technologies. [106].

4.10 Solar Thermal Installations

Concentrated solar thermal power (CSP, also known as concentrated solar power, or solar thermal power) generates solar energy by using mirrors or lenses to concentrate a large area of sunlight onto a receiver. CSP has a high potential in supplying renewable electricity on demand not only to the south but also to central and northern European countries, complementing wind and photovoltaics.

Electricity is generated when concentrated light is converted into heat (solar thermal energy), which drives a heat engine (usually a steam turbine) connected to an electric power generator or powers a thermochemical reaction.

Concentrated solar thermal power had a total global installed capacity of 6451 MW in 2019, 382 MW more than in 2018. Spain accounted for almost half of global thermo capacity, with 2300 MW, despite the fact that no new capacity has entered commercial operation in the country since 2013 [103].

Because both use energy from the sun, solar thermal is often compared to solar photovoltaics (PV), which experienced strong growth in recent years due to falling panel manufacturing prices and modular growth.

However, there are notable differences, resulting in solar thermal being in a competitive market against gas for generation at night, rather than competing against PV during the day. Solar thermal uses a different form of insolation than PV, direct normal irradiance (DNI), which is best found in desert regions, while PV uses global horizontal irradiance (GHI), which is more widely available globally, even at higher latitudes.

As a thermal power plant, solar thermal has more in common with thermal power plants such as coal or gas or geothermal energy. A solar thermal power plant can incorporate thermal energy storage, which stores energy in the form of sensible heat or as latent heat (e.g. using molten salts), allowing these plants to continue generating electricity as needed, day or night, unlike wind and photovoltaics. This makes solar thermal a dispatchable form of solar energy. Dispatchable renewable energy is particularly valuable in places where there is already high penetration of photovoltaics (PV) because an evening peak is created as PV drops at dusk (a phenomenon called the duck curve).

Solar thermal accounted for less than 2% of the world's installed capacity of solar power plants. However, in recent years, falling solar thermal prices are making the technology competitive with other baseload power plants that use fossil and nuclear fuels even in high humidity and dusty environments at sea level, such as the United Arab Emirates.

2015 was a year of challenges and changes for solar thermal power. Global operational capacity increased by 420 MW to reach around 4.8 GW by the end of the year. The year was a turning point in the expansion of a market that, to date, has been a game between two players: Spain and the United States account for almost 90% of installed capacity. But things are changing, with several facilities under construction by the end of 2015 in Australia, Chile, China, India, Israel, Mexico, Saudi Arabia, and South Africa. In fact, in 2015, Morocco and South Africa surpassed the United States in new capacity additions.

Nevertheless, Spain and the United States lead this market, with Spain leading the way with a cumulative capacity of 2.3 GW, followed by the United States with 1.7 GW.

Extremadura Solaben Solar Platform. 200 MW. Spain

The 200 MW Extremadura Solar Platform is the largest solar thermal complex in Europe and one of the largest plants of its kind in the world, located in the municipality of Logrosán, in the province of Cáceres, Extremadura, Spain. The complex was built in two phases and consists of four concentrated solar power (CSP) plants, Solaben 1, 2, 3, and 6, with an installed capacity of 50 MW each. Solaben 2 and 3, the two units belonging to the first phase, started operations respectively in July and December 2012, while the two units in the second phase, Solaben 1 and 6, started commercial operations in September 2013.

The two CSP plants in the first phase are jointly owned 70% by Abengoa Solar and 30% by Itochu, while the other two are wholly owned by Abengoa Solar. The construction cost of the first phase exceeded 500 million euro, 340 million euro of which was financed mainly through a loan from SMBC, HSBC, Mizuho, BTMU, and the Japanese export credit agency NEXI, while the second phase of the project was carried out with a total investment of 200 million euro. The implementation of the project has created 3,000 jobs during construction and 91 permanent jobs for its operation.

Andasol Solar Power Plant. 150 MW. Spain

The Andasol Solar Power Plant was built in Guadix, Andalusia, Spain. It was the first solar power plant in Europe to use a parabolic trough solar concentrator for commercial purposes. The plant consists of three installations with a capacity of 50 MW each, generating a total of approximately 540 GWh per year.

Each unit of the Andasol power plant consists of 312 rows of collectors made up of 28 mirrors and three absorber tubes, with molten salt used to store the solar thermal heat. The solar energy is stored in tanks to generate electricity even during the night. The Andasol 1 and Andasol 2 installations are owned by Solar Millennium and ACS Cobra, while Andasol 3 is owned by Marquesado Solar.

Solnova Solar Power Plant. 150 MW. Spain

The 150 MW Solnova Solar Power Plant is located in Sanlúcar la Mayor, Spain, forming part of the Solúcar Platform comprising five separate units of 50 MW each, of which only three units are currently in operation.

The plant uses ASTRØ parabolic trough technology, which involves the use of curved heliostatic mirrors to reflect sunlight onto a tube containing a fluid (synthetic oil). This fluid is heated to a temperature of 400 °C by solar radiation and used to produce steam that drives a turbogenerator to produce electricity. Abengoa Solar is the owner and operator of the Solnova Solar Power Plant, whose three installed units cost 710 million euros to build.

Extresol Solar Power Plant. 150 MW. Spain

The Extresol Solar Power Plant is located in Badajoz, Spain, which consists of three 50 MW units built in three phases, with a generation capacity of 175 GWh per year. The first unit began operation in 2010, while construction of the final unit was completed in August 2012.

The facility, which spans an area of approximately 500,000 m^2, uses parabolic trough technology employing molten salt to store solar heat, which each unit can store for 7.5 h. The energy produced at the plant, owned by ACS/Cobra Group, is distributed by Endesa.

4.10.1 Solar Heating and Cooling (SHC)

Solar thermal installations are used to produce domestic hot water in residential and industrial buildings. The main challenge of these solar-driven installations is to integrate this technology into the heating and cooling systems of the buildings. Regarding the economic aspect, the viability of Solar Heating and Cooling (SHC) is heavily dependent on local conditions and the price of the fuel used in the conventional system that SHC would replace. There is enough solar resource to use this technology in almost all European regions. The European Union supports the development of reliable and economically viable solar-driven solutions for the generation of domestic hot water, district heating, and industrial and commercial sites through research and innovation [107].

In 2017 an important milestone was reached regarding the solar heating and cooling sector. The European solar heat capacity reached 35 GWth of total installed capacity after a growth of 2.1% for solar thermal and an additional 1.4 GWth of solar collectors installed [108]. This installed capacity generated 25 TWh of heat in Europe. This means an area of about 50 million m2 of solar thermal collectors in Europe. This capacity is estimated to have a generation capacity of 25 TWh, which also means that there are more than 10 million households in Europe using solar thermal technology. These values are still far from the objective of the Member States for solar heat in 2020 of 78 TWh, meaning that the potential of the solar thermal technology in Europe is still unexploited. The International Renewable Energy Agency (IRENA) estimated that 3% of the European heat demand could be provided by solar thermal technologies by 2030 [103].

Greece was pioneer in the development and installation of solar thermal systems, their market grew 16% in 2017. This was due to the growth in competition among the thermosiphon system market, which brought down the prices in the market and made it more attractive to consumers. In Italy, the thermosiphon market also increased around 15–20%. The users reported that this technology works better than forced circulation.

In other countries, such as Denmark for example, the installed capacity of the solar thermal technology went from installing 335 MWth in 2016 to just 22 MWth in 2017. This was mainly caused by stopping the support in the investment in renewables and energy efficiency while the government was reassessing its policy. When a new framework was in place, the market in Denmark started to recover in 2018.

Solar district heating is slowly finding its place in other European countries as for example Austria, Germany, and France. The most relevant project regarding this technology is the flagship project "Big Golar Graz" a solar district heating system of around 200 MWth [109].

The solar thermal market grew 1.5% in 2019 in the EU, being the total installed area at just under 2.3 million m2. Each country's situation is different regarding solar thermal technology; therefore, this sector needs to reinvent itself to meet the ambitious challenge of climate neutrality [110].

4.11 Conclusions

The European Union faces several challenges related to the field of energy. These challenges include an increasing dependence on imports, limited diversification, high and unstable prices of energy, rising global power demand, risks of security affecting producer and transit regions, growing threats arising from climate change, and slow development in energy efficiency.

The European Green Deal, the ambitious EU climate policy, has the objective to become Europe the world's first climate-neutral continent by 2050. The measures taken related to this deal have the goal to enable European citizens and businesses to benefit from sustainable green transition. For this purpose, it is essential to transform the European energy system and boost the integration of renewable energy sources into energy systems.

The renewable power consumption augmented from 9.6% in 2004 to 18.9% in 2018. For the first time, in the first half of 2020, the EU generated more electricity from renewable sources than from fossil fuels. Wind, solar, hydro, and bioenergy produced 40% of the electricity among the member states, while fossil fuels generated 34%.

Wind power has been one of the most important energy sources in this success and it will be indispensable to reach the EU's renewable energy goals and achieve carbon neutrality by 2050. As the public acceptance gets higher, offshore wind energy signifies a growing future opportunity since the resources are stable and abundant. Research and innovation are also key factors in lowering costs and increasing the performance and reliability of offshore wind technology. It can significantly contribute to growing the share of renewable energy in Europe's mix. However, there are several challenges that need to be taken into accounts, such as the competing uses of the sea, costs, and grid integration, as well as improvements on the potential impacts of offshore turbines on the environment and public health.

Regarding solar energy, in 2019 Europe installed 16.7 GW, an increase of 104% compared to the 8.2 GW installed the previous year. Moreover, this year also saw the highest solar growth since 2010, when the European PV market also increased by 104% during the first solar boom, albeit at a lower level, reaching 13.4 GW.

References

1. Aseev AA, Blagovolin NS, Serebryannyi LR (1984) Exogenic landforms of Europe. In: Embleton C (ed) Geomorphology of Europe. Palgrave, London. https://doi.org/10.1007/978-1-349-17346-4_3
2. Magocsi PR (2002) Historical atlas of Central Europe (ed). University of Toronto Press, p 274. ISBN 0802084869, 9780802084866
3. Tockner K, Uehlinger U, Robinson CT (2009) Rivers of Europe (ed). Academic Press, p 728. ISBN0080919081, 9780080919089
4. Tockner K, Uehlinger U, Robinson CT, Tonolla D, Siber R, Peter FD (2009) European rivers. In: Likens GE (ed) Encyclopedia of inland waters, 1st edn. Elsevier, Academic Press, Oxford, pp 366–377
5. Niedźwiedź T, Obrębska-Starklowa B (1991) Klimat. In: Dynowska I, Maciejewski M (eds) Dorzecze górnej Wisły, PWN, Warszawa-Kraków, pp 68–84
6. Delanty G (2013) The historical regions of Europe: civilizational backgrounds and multiple routes to modernity. In: Formations of European modernity. Palgrave Macmillan, London. https://doi.org/10.1057/9781137287922_11
7. Zabecki DT (ed) World War II in Europe: an encyclopedia, p 1550. Military History of the United States, Routledge. ISBN 113581242X, 9781135812423
8. Horne J (ed) (1997) State, society and mobilization in Europe during the First World War. Volume 3 de studies in the social and cultural history of modern warfare, 312 pages. Cambridge University Press. ISBN0521561124, 9780521561129
9. Young JW (1994) Cold War Europe, 1945–1989: a political history. English Historical Rev 109(433):1039. Accessed 14 Feb 2021
10. Fink CK (2017) Cold war: an international history, 2nd edn. Westview Press, Boulder, CO
11. https://europa.eu/european-union/about-eu/history_en
12. Eichengreen B (1996) Institutions and economic growth: Europe since World War II. In: Crafts NFR, Toniolo G (eds) Economic growth in Europe since 1945. Cambridge University Press, Cambridge, pp 38–72
13. Jobs RI (2009) Youth movements: travel, protest, and Europe in 1968. Am Hist Rev 114:376–404
14. Bermann G (1989) The single European Act: a new constitution for the community? Columbia J Transnational Law 27:529–587
15. Van Esch F (2012) Why Germany wanted EMU: the role of Helmut Kohl's belief system and the Fall of the Berlin Wall. German Polit 21(1):34–52.https://doi.org/10.1080/09644008.2012.655012
16. Harrison H (2003) Driving the Soviets up the wall: Soviet-East German Relations, 1953–1961. Princeton University Press, Princeton, NJ
17. Navarro MF, Fabero, España. La división de Alemania y el Muro de Berlín (1949–1961), Agosto 2012. ISSN 1853-2756. www.huellasdelahistoria.com
18. Baun MJ (1995) The Maastricht Treaty as high politics. Polit Sci Q 110(4):605–624
19. Baun MJ (1996) An imperfect union: The Maastricht Treaty and the New Politic of European Integration. Westview, Boulder, CO/Oxford
20. Moravcsick A, Nicola'idis K (1999) Explaining the Treaty of Amsterdam: interest, influence, institutions. J Common Mark Stud 37:59–85

21. Manin P (1998) The Treaty of Amsterdam. Columbia J Eur Law 4:1–68
22. Havlíček T, Jeřábek M, Dokoupil J (eds) (2018) Borders in Central Europe after the schengen agreement. Springer International Publishing
23. Whitaker E (1992) The Schengen agreement and its portent for the freedom of personal movement in Europe. Georgetown Immigration Law ournal 6:191–222
24. Antonakakis N, Chatziantoniou I, Gabauer D (2021) The impact of Euro through time: exchange rate dynamics under different regimes. Int J Financ Econ 26(1):1375–1408
25. Gibson HD, Palivos T, Tavlas GS (2014) The crisis in the Euro Area: an analytic overview. J Macroecon 39:233–239
26. Pisani-Ferry J, Sapir A, Veron N, Wolff G (2012) What kind of European BankingUnion? Policy Contribution 2012/12. Bruegel, Brussels
27. Fijnaut C (2015) Revolution or evolution through the Treaty of Lisbon: police cooperation in Europe in a broader historical context. In: Aden H (ed) Police cooperation in the European Union under the Treaty of Lisbon: opportunities and limitations. Nomos, Baden-Baden, pp 25–48
28. Pernice I (2008) The Treaty of Lisbon and fundamental rights. Walter Hallstein-Institut für Europäisches Verfassungsrecht. WHI - Paper 7/08
29. Bercusson B (2009) The Lisbon Treaty and Social Europe, ERA Forum, 2009, 10, 87-105
30. Filip A, Raffournier B (2012) The impact of the 2008–2009 financial crisis on earnings management: The European evidence. In: Illinois international journal of accounting symposium, Ontario, Canada, July 5–7
31. Alkopher TD, Blanc E (2016) Schengen area shaken: the impact of immigration-related threat perceptions on the European security community. J Int Relat Dev. https://doi.org/10.1057/s41268-016-0005-9
32. Bauböck R (2017) Europe's Commitments and failures in the refugee crisis. Eur Polit Sci, pp 1–11. https://doi.org/10.1057/s41304-017-0120-0
33. https://ec.europa.eu/info/european-union-and-united-kingdom-forging-new-partnership/brexit-brief/transition-period_es
34. Morris M, Kibasi T (2018) The Brexit withdrawal agreement. Institute for Public Policy Research
35. Prescott C, Pilato M, Bellia C (2020) Geographical indications in the UK after Brexit: an uncertain future? Food Policy 90:101808
36. Böröcz J, Sarkar M (2005) What Is the EU? Int Sociol 20(2):153–173
37. https://europa.eu/european-union/about-eu/figures/economy_es
38. https://europa.eu/european-union/about-eu/figures/living_es
39. Lindh T, Malmberg B (2009) European union economic growth and the age structure of the population. Econ Change Restruct 42(3):159–187
40. Anghelache C, Marinescu AI, Samson T (2018) Theoretical characteristics of the purchasing power parity in the EU context. Int J Acad Res Account Fin Manage Sci 8(2):37–47
41. Rechel B, Grundy E, Robine JM, Cylus J, Mackenbach JP, Knai C, McKee M (2013) Ageing in the European Union. Lancet 381:1312–1322. https://doi.org/10.1016/S0140-6736(12)62087-X
42. Pasimeni F, Pasimeni P (2016) An institutional analysis of the Europe 2020 strategy. Soc Indic Res 127:1021–1038. https://doi.org/10.1007/s11205-015-1013-7
43. Rogge N (2019) EU countries' progress towards 'Europe 2020 strategy targets.' J Policy Model 41:255–272
44. Stec M, Grzebyk M (2018) The Implementation of the strategy Europe 2020 objectives in European Union Countries: the concept analysis and statistical evaluation. Qual Quant 52:119–133
45. Syssner J (2020) Policy implications of rural depopulation. In: Pathways to demographic. Springer: Berlin/Heidelberg, Germany, pp 37–52
46. Viñas CD (2019) Depopulation processes in European rural areas: a case study of Cantabria (Spain). Eur Countrys 11:341–369

47. Rontos K, Colantoni A, Salvati L, Mosconi EM, Morera AG (2020) Resident or present? Population census data tell you more about suburbanization. Land 9:383. https://doi.org/10.3390/land9100383
48. Goos M, Manning A, Salomons A (2009) Job polarization in Europe. Am Econ Rev 99(2):58–63
49. https://ec.europa.eu/regional_policy/sources/docgener/studies/pdf/citiesoftomorrow/citiesoftomorrow_final_es.pdf
50. Ruble I (2017) European Union energy supply security: the benefits of natural gas imports from the Eastern Mediterranean. Energy Policy 105:341–353
51. https://elordenmundial.com/mapas/dependencia-energetica-europa/
52. https://datosmacro.expansion.com/energia-y-medio-ambiente/petroleo/reservas
53. https://www.miteco.gob.es/eu/calidad-y-evaluacion-ambiental/publicaciones/POTENCIAL%20ENERGIA%20EOLICA%20TERRESTRES%20Y%20MARINA_tcm35-185588.pdf
54. Miglietta MM, Huld T, Monforti-Ferrario F (2017) Local complementarity of wind and solar energy resources over Europe: an assessment study from a meteorological perspective. J Appl Meteorol Climatol 56(1):217–234
55. Etxebarria Berrizbeitia S (2017) Sky-diffuse radiation models for the globe, Universidad de Granada. ISBN: 9788491634744, http://hdl.handle.net/10481/48219
56. OECD International Energy Agency (2020) World energy outlook 2020
57. World Nuclear Performance Report (2020) Produced by: World Nuclear Association. Published: August 2020, Report No. 2020/008. https://www.world-nuclear.org/getmedia/3418bf4a-5891-4ba1-b6c2-d3d8907264d/performance-report-2020-v1.pdf.aspx
58. https://ec.europa.eu/eurostat/statistics-explained/index.php/Renewable_energy_statistics#Share_of_renewable_energy_almost_doubled_between_2004_and_2018
59. Proskurina S, Sikkema R, Heinimö J, Vakkilainen E (2016) Five years left–How are the EU member states contributing to the 20% target for EU's renewable energy consumption; the role of woody biomass. Biomass Bioenerg 95:64–77
60. Jones D, Moore C (2020) Renewables beat fossil fuels. A half-yearly analysis of Europe's electricity transition, July 22, 2020. https://ember-climate.org/project/renewables-beat-fossil-fuels/
61. IRENA (2020) Renewable power generation costs in 2019, International Renewable Energy Agency, Abu Dhabi
62. Almuni M, Dauwe T, Moorkens I, Saarikivi RJ, Tomescu MS (EEA) Eionet report—ETC/CME 7/2020—December 2020 Renewable energy in Europe—2020. Recent growth and knock-on effects
63. https://ec.europa.eu/eurostat/databrowser/view/nrg_ind_ren/default/table?lang=en
64. https://ec.europa.eu/eurostat/web/products-eurostat-news/-/DDN-20200129-1
65. https://ec.europa.eu/info/research-and-innovation/research-area/energy-research-and-innovation/wind-energy_en
66. https://ec.europa.eu/eurostat/cache/infographs/energy/bloc-4a.html
67. https://www.europarl.europa.eu/factsheets/es/sheet/68/la-politica-energetica-principios-generales
68. https://www.consilium.europa.eu/media/24561/145397.pdf
69. https://ec.europa.eu/info/strategy/priorities-2019-2024/european-green-deal/delivering-european-green-deal_en
70. https://ec.europa.eu/info/news/focus-renewable-energy-europe-2020-mar-18_es
71. https://ec.europa.eu/eurostat/cache/infographs/energy/bloc-1.html
72. https://ec.europa.eu/info/policies/energy_es
73. https://ec.europa.eu/energy/topics/energy-strategy-and-energy-union_es
74. https://ec.europa.eu/energy/topics/energy-strategy/clean-energy-all-europeans_en
75. https://ec.europa.eu/energy/topics/renewable-energy/onshore-and-offshore-wind_en
76. https://windeurope.org/about-us/new-identity/

77. https://windeurope.org/wp-content/uploads/files/about-wind/statistics/WindEurope-Annual-Statistics-2019.pdf
78. Bórawski P, Bełdycka-Bórawska A, Jankowski KJ, Dubis B, Dunn JW (2020) Development of wind energy market in the European Union. Renewable Energy 161:691–700
79. Bilgili M, Yasar A, Simsek E (2011) Offshore wind power development in Europe and its comparison with onshore counterpart. Renew Sustain Energy Rev 15(2):905–915
80. Komusanac I, Ramirez L, Brindley G, Fraile D (2020) The impact of COVID-19 on Europe's wind sector. Executive summary
81. Hosseini SE (2020) An outlook on the global development of renewable and sustainable energy at the time of COVID-19. Energy Res Soc Sci 68:101633
82. https://www.windpower.by/en/news/1061.html
83. https://energetica21.com/noticia/-la-resistente-industria-eolica-suministra-el-17porciento-de-la-electricidad-en-europa
84. Windfair Editors. Turbine Orders Surge in 2020 Despite Pandemic. 01/21/2021. https://w3.windfair.net/wind-energy/news/36569-windeurope-turbine-order-2020-pandemic-offshore-onshore-wind-power-surge-increase-gw-uk-weden-netherlands-germany-france
85. https://windeurope.org/newsroom/news/record-15-gw-of-new-wind-turbines-ordered-in-2020/
86. https://windeurope.org/wp-content/uploads/files/about-wind/reports/Unleashing-Europes-offshore-wind-potential.pdf
87. Craig R, Ocean winds secures Polish offshore wind support Poland's energy regulatory office latest award of CfDs means it has now contracted just over 5.9 GW of offshore wind in the Baltic Sea. https://www.windpowermonthly.com/article/1720907/ocean-winds-secures-polish-offshore-wind-support
88. Campoccia A, Dusonchet L, Telaretti E, Zizzo G (2009) Comparative analysis of different supporting measures for the production of electrical energy by solar PV and Wind systems: four representative European cases. Sol Energy 83(3):287–297
89. Auer J, Keseric N, Haas R, Resch G, Huber C., & Faber, T. (2005, March). Medium-and long-term effects of EU-electricity enlargement. In Paper presented at the 3rd SESSA Conference on Perspectives and Challenges of EU Electricity Enlargement, Berlin. In proceedings.
90. Capizzi F, Das A, Dauwe T, Moorkens I, Juhana R, Tomescu M (2019) Renewable energy in Europe—2019. Mol, Belgium, European Environmental Agency-European Topic Centre on Climate Change Mitigation and Energy
91. Antolini A, Maritime issues—renewable energy including offshore wind power generation: The EU's Baltic Sea countries sign declaration of intent on the offshore wind development in the Baltic Sea. JTTRI/Japan Transport and Tourism Research Institute
92. Onea F, Rusu L (2018) Evaluation of some state-of-the-art wind technologies in the nearshore of the black sea. Energies 11(9):2452
93. Energy Policy Group (2020) Romania's offshore wind energy resources: natural potential, regulatory framework, and development prospects
94. https://windeurope.org/newsroom/significant-developments-on-offshore-wind-in-the-baltic-sea/
95. Foxwell D (2020) Europe's installed floating offshore wind capacity could hit 7 GW by 2030, trade group says. Institute for Energy Economics and Financial Analysis, 10 Sept 2020. https://ieefa.org/europes-installed-floating-offshore-wind-capacity-could-hit-7gw-by-2030-trade-group-says/
96. https://windeurope.org/newsroom/news/offshore-wind-is-coming-to-greece/
97. https://windeurope.org/newsroom/news/poland-adopts-historic-offshore-wind-act/
98. https://www.energias-renovables.com/fotovoltaica/espana-acaparo-en-2019-el-90-de-20201109
99. https://ec.europa.eu/energy/topics/renewable-energy/solar-power_en
100. EU Market Outlook. For Solar Power/2019–2023. Join 200+. Solar Power Europe Members. https://www.solarpowereurope.org/wp-content/uploads/2019/12/SolarPower-Europe_EU-Market-Outlook-for-Solar-Power-2019-2023_.pdf?cf_id=5387

101. Agora Energiewende and Sandbag (2020) The European power sector in 2019: up-to-date analysis on the electricity transition. https://ember-climate.org/wp-content/uploads/2020/02/Sandbag-European-Power-Sector-Review-2019.pdf
102. https://www.solarpowereurope.org/wp-content/uploads/2019/12/SolarPower-Europe_EU-Market-Outlook-for-Solar-Power-2019-2023_.pdf?cf_id=10355
103. IRENA (2019) Future of solar photovoltaic: deployment, investment, technology, grid integration and socio-economic aspects (A global energy transformation: paper). International Renewable Energy Agency, Abu Dhabi
104. Bódis K, Kougias I, Jäger-Waldau A, Taylor N, Szabó S (2019) A high-resolution geospatial assessment of the rooftop solar photovoltaic potential in the European Union. Renew Sustain Energy Rev 114:109309
105. Dodd N, Espinosa N (2021) Solar photovoltaic modules, inverters and systems: options and feasibility of EU Ecolabel and green public procurement criteria. Preliminary report, EUR 30474 EN, Publications Office of the European Union, Luxembourg. https://doi.org/10.2760/29743. ISBN 978-92-76-26819-2, JRC122430
106. Zhao ZY, Chen YL, Thomson JD (2017) Levelized cost of energy modeling for concentrated solar power projects: a China study. Energy 120:117–127
107. https://ec.europa.eu/info/research-and-innovation/research-area/energy-research-and-innovation/solar-energy_en
108. Solar Heat Markets in Europe Trends and Market Statistics (2017) Summary (November 2018). http://solarheateurope.eu/wp-content/uploads/2018/12/Solar_Heat_Markets_2018-Web-based-version.pdf
109. http://solarheateurope.eu/wp-content/uploads/2018/12/Solar_Heat_Markets_2018-Web-based-version.pdf
110. https://www.eurobserv-er.org/solar-thermal-and-concentrated-solar-power-barometer-2020/

Chapter 5
Wind Energy and Solar PV Developments in Japan

5.1 Introduction

Under the Paris Agreement, the Japanese government has set a target to reduce greenhouse gas emissions by 80% by 2050. Photovoltaic systems and wind power are expected to play an important role in achieving this target. In 2018, the Ministry of Economy, Trade, and Industry (METI) stated that the country's medium- and long-term energy policy, the "Strategic Energy Plan", should expand PV systems and wind power installations to account for the majority of energy supply [1–3].

As an island nation with limited natural resource endowments and without international gas pipelines and electricity connections, Japan faces fundamental energy security challenges.

Reliance on imported solid fuels soared to 94% of energy supply in 2014, but the resumption of nuclear power, expansion of renewables, and declining energy demand helped reduce this share to 88% in 2019.

Japan has successfully diversified its sources of liquefied natural gas (LNG) imports, while oil imports remain highly concentrated with a small number of suppliers in the Middle East. At the same time, Japan has one of the world's largest oil reserves, which acts as insurance against geopolitical risks and major global crises.

Achieving carbon neutrality by 2050 requires a sharp reduction in emissions as soon as possible and at the latest from 2030 onwards, and the rapid implementation of a comprehensive set of policies and measures. Japan's energy policy is guided by the principles of energy security, economic efficiency, environmental sustainability, and safety. The 5th Strategic Energy Plan, approved in 2018, aims to achieve a more diversified energy mix by 2030, with a larger share for renewables and the resumption of nuclear power. It also aims to improve the efficiency of fossil fuel use and reduce energy demand. In the 5th Strategic Energy Plans, the share of renewable energy in the Total Primary Energy Supply (TPES) is expected to reach 13% in 2030, up from 8% in 2019. Renewable energy generation is expected to reach 24% in 2030, up from 19% in 2019.

T. Muneer et al., *The Coming of Age of Solar and Wind Power*, Green Energy and Technology, https://doi.org/10.1007/978-3-030-92010-4_5

Japan has experienced a rapid expansion of solar PV in recent years, driven by generous feed-in tariffs. However, more efforts are needed to develop other renewable technologies, such as wind and geothermal, for which Japan has large energy potential.

In the energy sector, the main challenge is to address grid constraints, including through better connectivity between Japan's regional grids. Regulatory reform can help to improve the operation of the electricity system, thus enabling the integration of a higher share of variable renewable energy. The expansion of renewables in the heat and transport sectors remains slow due to weak political support.

Although fossil fuel use is expected to decrease, it remains high at 76% of TPES and more than half of power generation in 2030. The projected energy mix in 2030 is consistent with the target of reducing greenhouse gas emissions by 26% by 2030 compared to 2013 levels. However, in light of the recently announced ambition to become carbon neutral by 2050, the share of zero-emission energy sources needs to be increased.

5.2 Geographical Background

The Japanese territory is located in the region known as Northeast Asia or East Asia, east of the Eurasian continent. It is surrounded by the Pacific Ocean, the Sea of Okhotsk, the Sea of Japan, and the East China Sea.

Japan's easternmost tip is the island of Minami-Tori Shima (153° 59′ 12″ east longitude), its westernmost tip is the island of Yonagunijima (122° 55′ 57″ east longitude), its southernmost tip is the island of Oki-no-Tori Shima (20° 25′ 31″ north latitude) and its northernmost tip is the island of Etorofu (45° 33′ 26″ north latitude).

Japan is an archipelago of 377,974 km^2 of islands. Honshu, the largest island of approximately 228,000 km^2, is the 7th largest island in the world.

The capital of Japan is Tokyo and is located on the island of Honshu.

Approximately 84% of the territory is mountainous, 14% of the area is devoted to agricultural and livestock activities, 66% to forests, and the remaining 20% to other uses. Only about 25% of the territory is flat, where the population is concentrated.

The country is located at the convergence and subduction of several highly active tectonic plates such as the Philippine, Pacific, Eurasian, and North American plates. Its location on the Pacific Ring of Fire has meant that Japan has experienced five major earthquakes in the last 15 years.

In addition, there are 188 volcanoes throughout Japan. The best known is Mount Fuji, which is 3776 m high.

Most areas have a temperate climate with four distinct seasons, although Okinawa in the south is subtropical and Hokkaido in the north is subarctic. As a result, Japan has a great diversity of flora and fauna. There is also a wealth of valuable ecosystems throughout the country, with four areas on the UNESCO World Heritage List, Shiretoko, Shirakami-Sanchi, and the islands of Ogasawara and Yakushima.

At the same time, rainfall caused by monsoon winds blowing in from the Asian mainland makes Japan a rainy country in spring and autumn.

Japan's flora is varied and abundant, with more than 17,000 species, including Japanese chestnut, thuja, red pine, Korean pine, birch, and ash. There are also white and red plum trees and bamboo plants.

Japan's fauna includes about 132 species of mammals, 583 species of birds, 66 species of reptiles, fish, and batrachians, including the Asiatic black bear, brown bear, wild boar, deer, antelope, badger, raccoon dog, foxes, wild cat, sea lion, Japanese wolf, copper and green pheasant, and the Asiatic goshawk eagle.

5.3 Historical Background

Japan was one of the most isolated and traditionalist societies in the world when Commodore Matthew C. Perry entered Tokyo Bay in command of an American fleet in 1853. His aggressive foreign policy in the Pacific led Japan to go to war with the United States in 1941. The subsequent defeat ended Japan's dream of dominating Asia, and the US occupation imposed a constitutional parliamentary system, freedom of trade unions, and sweeping land reform. Despite a shortage of raw materials, the economy was revitalised by US aid, high rates of labour productivity, private savings, and capital investment.

The death of Emperor Hirohito in 1989 ushered in a new era in which Japan faces the challenges of an ageing population, an increasingly unequal distribution of wealth, the changing role of women in society, and growing concerns about security and the environment. Despite all this, the Japanese economy has experienced steady growth since the financial crisis of the 1990s. Japan has become a global driving force, establishing itself as the world's second largest economy.

Relations with North Korea are strained because of the country's nuclear weapons programme and the abductions of Japanese citizens in the 1970s and 1980s. Relations with Russia are hampered by the dispute over a series of small islands east of Hokaido, known as the Northern Territories: Habomai Shikotan, Kunashiri, and Etorofu. Japan continues to claim sovereignty over these islands occupied by Russia at the end of World War II.

Until 2008, Japan maintained only a single powerful political party, the Liberal Democratic Party, although minor parties existed. Since 2009, the strength of smaller political parties has grown, causing the power of the Liberal Democratic Party to diminish [4].

The current prime minister, Yoshihide Suga of the Liberal Democratic Party, took office in September 2020.

5.4 Socio-economics of Japan

In general, per capita, energy consumption is on the rise, for a variety of reasons, including increased access to and use of goods and services, mobility and communication requirements, aspirations for better living conditions, security, etc. It is, therefore, reasonable to assume that the higher a country's per capita energy consumption, the higher its quality of life and socio-economic development. Japan's per capita energy consumption for the year 2019 was 7,154.6 KWh, ranking 20th among the world's nations with the highest per capita consumption [5] (Fig. 5.1).

Japan is the 3rd largest economy in the world in terms of Gross Domestic Product (GDP) [6]. In 2019 the GDP figure was €4,525,204 million. After falling into recession following the Great East Japan Earthquake of 2011 and the subsequent Fukushima Daiichi nuclear accident, GDP growth has fluctuated between 0.3% and 2.2% since 2012 [7]. Japan's GDP per capita in 2019 was €35,860, making it 26th (Fig. 5.2). Japan is the country with the highest public debt to GDP ratio. Specifically, the debt was 234.86% of its GDP. Its per capita debt was 85.591 euros per capita [8]. In contrast, it is the country with the lowest unemployment in the world. In 2021, Japan's unemployment rate was 2.6%, a negligible rate [9]. Due to containment measures and lower external demand caused by the COVID-19 pandemic, GDP will fall by 6–7%, with a modest recovery expected by 2021 [10].

Japan is among the most populous countries in the world, ranking eleventh, with a population of 126,529,100 at the end of 2020. It has a very high population density of 334 inhabitants per square km [11].

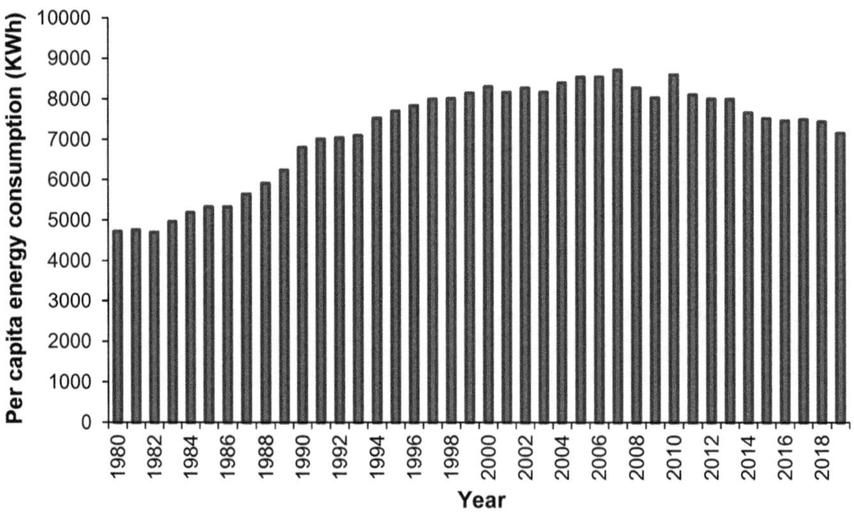

Fig. 5.1 Japan's per capita energy consumption (KWh). *Source* prepared and adapted by the author on the basis of https://data.worldbank.org/indicator/EG.USE.ELEC.KH.PC?locations=JP-CN

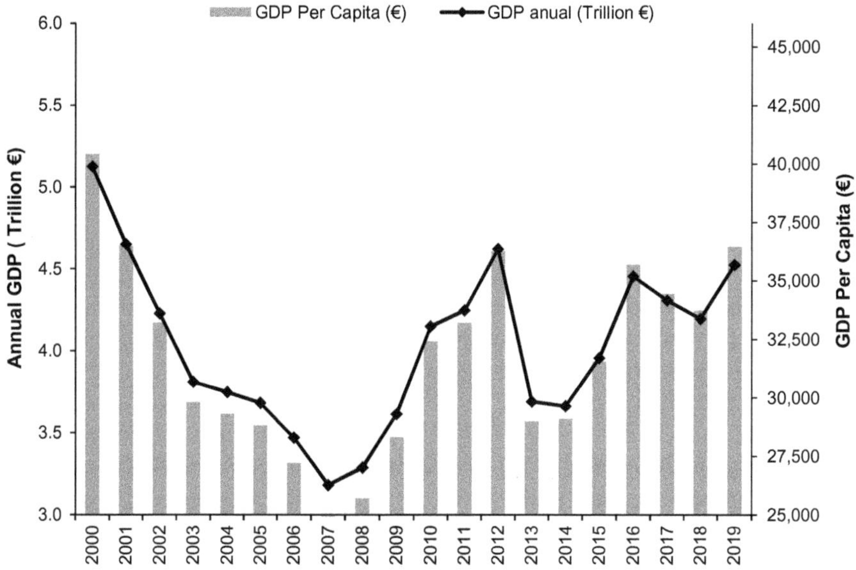

Fig. 5.2 Annual GDP and GDP Per Capita (2019). *Source* prepared and adapted by the author on the basis of https://data.worldbank.org/indicator/NY.GDP.MKTP.KD.ZG?locations=CN

From north to south Japan is divided into eight regions: Hokkaido, Tohuku, Kanto, Chubu, Kinki, Chugoku, Shikoku and Kyushu. Each of these regions is made up of prefectures, with the exception of the Hokkaidō region, which is made up of a single prefecture. Prefectures are the official administrative unit of the country and are commonly referred to as Todōfuken. There are 47 territorial jurisdictions (1 metropolitan district: Tokyo, 1 province: Hokkaidō, 2 urban prefectures: Osaka and Kyoto, and 43 rural prefectures). The capital of Japan, Tokyo, is by law a special administrative unit called to (metropolitan district) which has authority equivalent to that of a prefecture and is therefore listed as one of Japan's 47 prefectures. Its official name is "Tokyo Metropolis", and it has incorporated cities under its jurisdiction. Each of Tokyo's 23 special wards is legally a city.

Primary activities: Agriculture, forestry, fishing, and mining are only a small part of the Japanese economy. These are economic activities that, although they do not produce as much value as the other sectors, are highly developed. The country does not have much land suitable for agriculture, but the land used for this purpose is very productive. Only 20% of the country's land area is arable, and agricultural production is highly subsidised. Most agricultural production stays in the country, but not enough, which is why Japan imports about 60% of its food. Japan's main agricultural product is rice, with rice fields accounting for 55% of cultivated land. But it accounts for only one-fifth of the value of agricultural production. Other important crops are soybeans, barley, wheat, peanuts, and sugar beets. Fruit production is dominated by apples, pears, and citrus fruits. Livestock accounts for almost 30% of the value of Japanese primary production. In monetary terms, milk production is the most important activity, followed by pig farming, cattle farming, poultry farming, and forestry. The latter is carried out through advanced techniques in aquaculture and marine farming. Finally, forestry and mining are not very developed. The former because most forestry activities in the country are part-time activities of farmers and small companies. The latter is because of the scarcity of natural resources in the country. It is estimated that these primary activities account for 1.2% of GDP.

Secondary activities: They account for 27.5% of GDP [5]. The power of the Japanese economy is largely due to its dominant manufacturing industry. The country is recognised worldwide for the quality and innovation of its industrial products. This sector is highly dependent on imports of raw materials and fuels. In addition, Japan is renowned for producing high-value products. It transforms imported raw materials into high-value consumer goods for export.

Tertiary activities: The service sector is the most important sector of Japan's economy, accounting for 71.4% of GDP [5]. It is also the most dynamic sector, surpassing the industrial sector. This sector encompasses many economic activities; the most important are retail trade, advertising, tourism, entertainment, restaurants, education and health, television, and real estate. The main Japanese industries are the production of automobiles, consumer electronics, computers, and semiconductors. Other relevant sectors are petrochemicals, pharmaceuticals, textiles, and processed foods. One of the most successful industries is the automotive industry. Japan has a large market share in automobiles, tyres, electrical machinery, and engines.

5.5 Energy Resources

The events of 2011 (Great East Earthquake and the subsequent Fukushima Daiichi nuclear accident) have had a significant impact on Japan's energy system. The accident led to the total suspension of the nuclear fleet, which has only been partially restarted. This has made Japan more dependent on fossil fuels to close the gap in electricity generation [12].

Fossil fuels accounted for 661 TWh of Japan's net electricity generation in 2019, accounting for about 70% of total generation, up from 61% in 2010. However, the share of fossil fuels has been steadily declining since a peak of 85% in 2012 as a result of nuclear reactor restarts and the exponential growth of renewables.

The power sector (Fig. 5.3) uses coal as a base load source for power generation. The share of coal in the power sector was 26% before the Fukushima accident and rose to 31% in 2019 [13]. Natural gas, which accounted for 27% of Japan's electricity generation in 2010, increased its share to 34% in 2019 [14]. The use of fuel oil and crude oil in the power sector has declined since the Fukushima accident, accounting for 5% of generation in 2019, down from 17% of Japan's electricity generation in 2012 and below the levels seen before Fukushima [9]. Nuclear generation increased to 7% of Japan's electricity production slate in 2019, up from no production in 2014. Japan's government intends to boost production to 20–22% of total generation by 2030 [15]. Japan decommissioned all nuclear reactors following the Fukushima

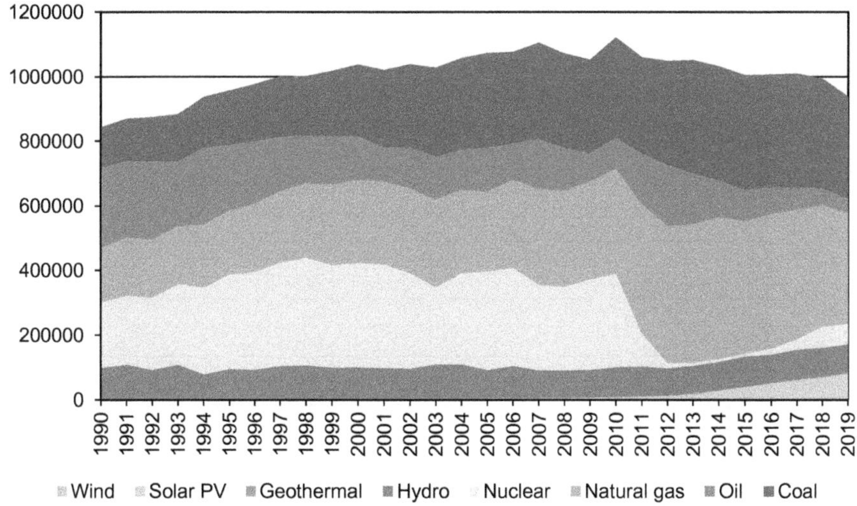

Fig. 5.3 Electricity generation by source (GWh), Japan 1990–2019. *Source* prepared and adapted by the author on the basis of www.iea.org/statistics

accident. The country's nuclear generation was reduced to zero in September 2013 and remained at zero for almost two years.

Like nuclear power, hydropower is a base load generation source in Japan due to low generation costs and relatively stable supply. Net hydropower generation accounted for about 8% of Japan's total net generation mix in 2019. Renewable energy, excluding hydro, accounted for 6% of Japan's total energy consumption and more than 15%, or 146 TWh, of the country's total electricity generation in 2019 [16]. Most of the growth in renewable capacity over the last decade came from solar power as a result of strong investment in large-scale solar PV units and financial incentives from the government's electricity buyback programme and the feed-in tariff (FIT) programme. By the end of 2019, Japan's solar generation capacity reached 62 GW, 10 times more than the capacity of 2012 [17]. Japan identified renewable energy as an important fuel source for long-term energy security and emissions reduction and intends to increase total renewable energy to 22–24% of total generation by 2030 [18].

5.5.1 Petroleum and Other Liquids

Japan has limited proven domestic oil reserves, totalling 44 million barrels as of January 2020 [19].

In 2019, Japanese production of oil and other liquids was about 127,000 barrels per day (b/d), with only about 10,000 b/d of that total coming from light crude oil and natural gas liquids (Fig. 5.4). Most of Japan's domestic liquids supply comes

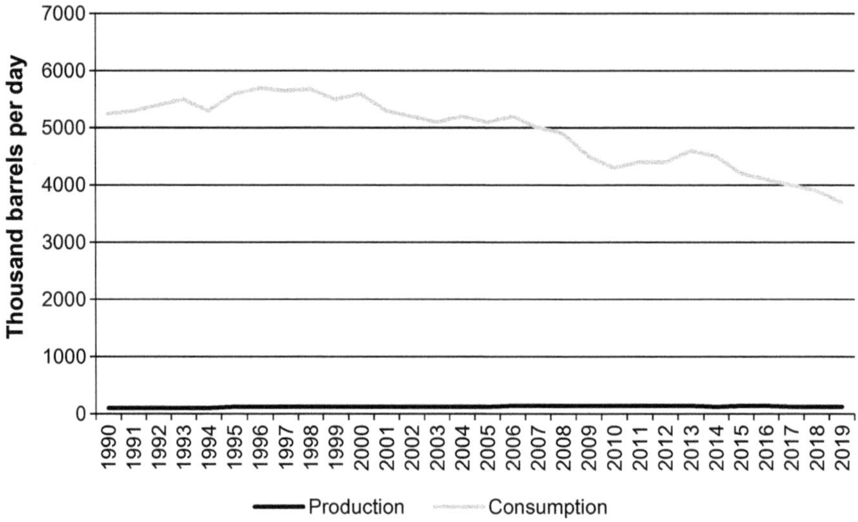

Fig. 5.4 Japan's petroleum and other liquids production and consumption, 1990–2019. *Source* prepared and adapted by the author on the basis of https://www.eia.gov/outlooks/steo/pdf/steo_full.pdf

from refineries because the country has a large oil refining sector. According to the International Energy Agency, Japan had about 388 million barrels of strategic crude oil reserves in June 2020. 76% of these stocks were government stocks and 24% were commercial stocks [20].

5.5.2 *Natural Gas*

Japan had 738 billion cubic feet (Bcf) of proven natural gas reserves in January 2020 [19]. In 2019, production was 76 Bcf, down from a peak of 139 Bcf in 2007 (Fig. 5.5). Domestic production accounted for only 2% of Japan's natural gas consumption [21].

5.5.3 *Coal*

Domestic coal production is negligible and Japan relies almost exclusively on imports. Coal imports increased to 183 Mt in 2019 from 175 Mt in 2011, after more coal-fired generation capacity came online. Japan was the world's third largest importer of coal in 2019, behind China and India [22]. Australia continues to provide the largest share of Japan's coal imports (60% in 2019), followed by Indonesia (15%), Russia (11%), and other countries (14%) [23]. Japan began diversifying its coal

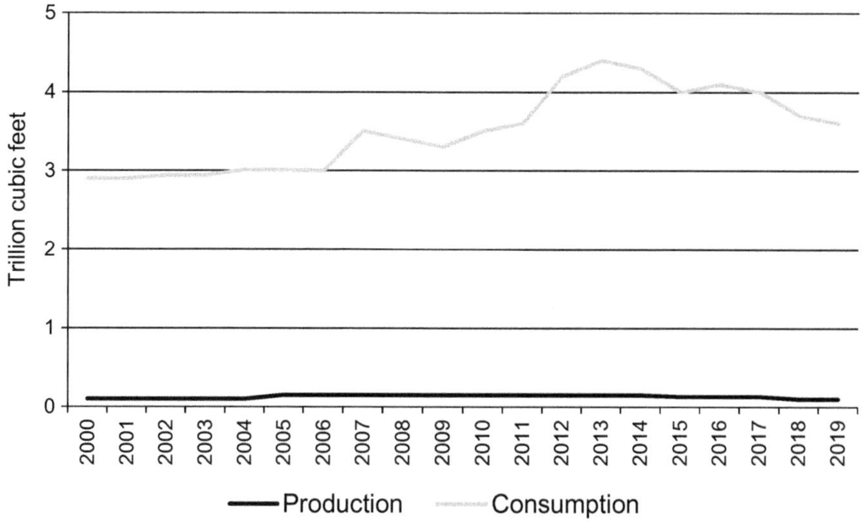

Fig. 5.5 Japan's dry natural gas production and consumption, 2000–2019. *Source* prepared and adapted by the author on the basis of https://www.eia.gov/outlooks/steo/pdf/steo_full.pdf

import sources in 2018, with the United States, Russia, and Canada taking larger shares of the global portfolio (Fig. 5.6).

5.6 Energy Demand

Energy consumption in Japan amounted to 283 Mtoe in 2018, representing 66% of TPES. Consumption has decreased by 10% since 2008. Energy demand rebounded slightly in 2017 but declined again in 2018.

Industry was the largest energy consuming sector, with 41% of TPES in 2018, followed by transport with 25%, services with 19%, and the residential sector with 15%. All sectors have seen their energy consumption gradually decline since 2008. After a drop in energy consumption during the global financial crisis 2007/08 (Fig. 5.7), energy consumption picked up again in 2010 with the economic recovery. The overall trend, however, has been steadily declining, with a 5% drop in the period 2019–2016.

In 2017, energy consumption increased again (+1.2%), driven by higher consumption in industry and the residential sector, but then fell again in 2018 to the lowest value since 1990. Following the Great East Japan Earthquake of 2011 and the subsequent nuclear accident at the Fukushima Daiichi power plant, the government launched a nationwide energy-saving movement called Setsuden to address the sudden drop in nuclear power generation. Households and businesses were asked to make major efforts to save electricity, for example by changing the hours and days of operation of

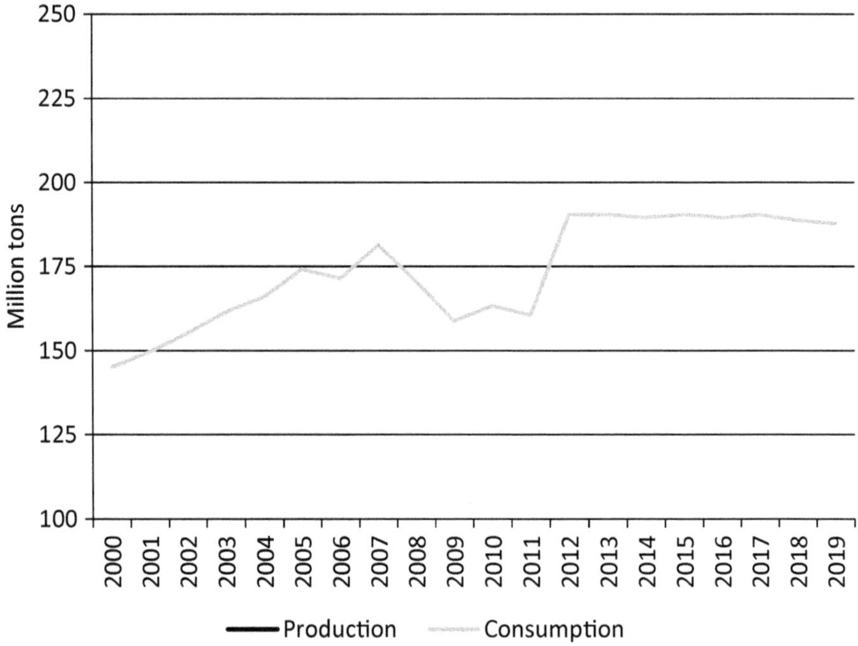

Fig. 5.6 Japan's coal production and consumption, 2000–2019. *Source* prepared and adapted by the author on the basis of https://www.eia.gov/outlooks/steo/pdf/steo_full.pdf

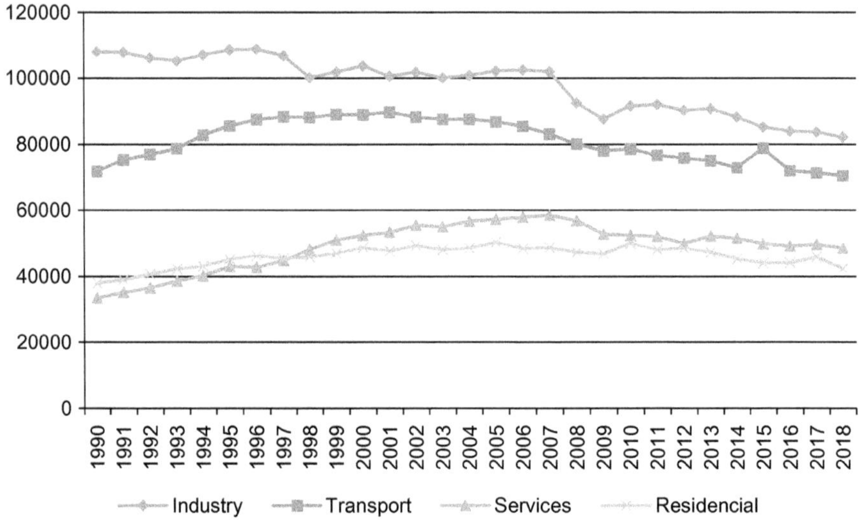

Fig. 5.7 Total final consumption (TFC) by sector, Japan, 1990–2018. *Source* prepared and adapted by the author on the basis of www.iea.org/statistics

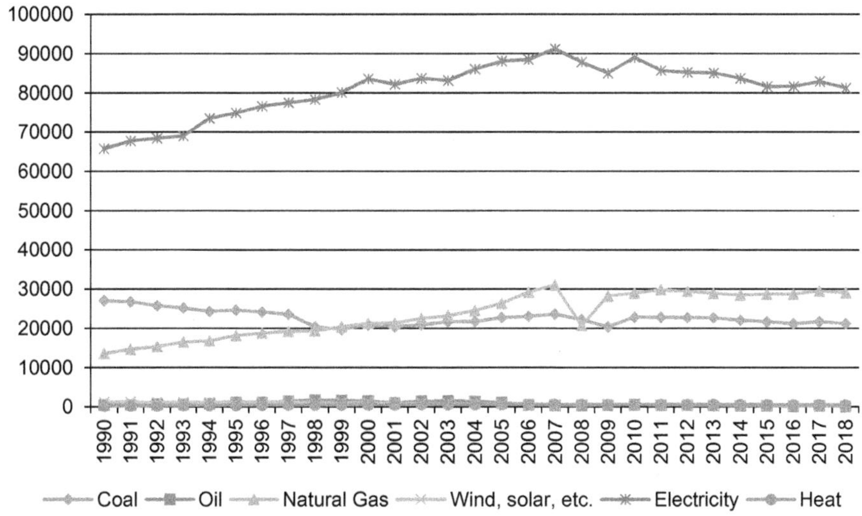

Fig. 5.8 Total final consumption (TFC) by source, Japan 1990–2018. *Source* prepared and adapted by the author on the basis of www.iea.org/statistics

industry, turning off lights, or limiting the use of air conditioning or electric heating during peak hours.

Oil accounted for more than half of TFC in 2018 and dominates in particular transport energy demand, at 97%. Oil is also the largest source of energy in industry, at 43% of total demand, and accounts for 26% and 27% in the residential and services sectors, respectively. Electricity is the second largest source of TFC, covering more than half of the energy demand of buildings in both the residential and service sectors (Fig. 5.8).

5.6.1 *Petroleum and Other Liquids*

Japan's oil consumption was 3.7 million b/d in 2019, making it the world's fifth-largest oil consumer behind the United States, China, India, and Russia.

However, oil demand in Japan fell by more than 1 million b/d between 2012 and 2019 (Fig. 5.4) [24]. Structural factors such as a shrinking and ageing population, high energy efficiency measures, and a growing fleet of hybrid and electric vehicles continue to reduce oil demand [25].

In 2019, Japan's liquid fuel consumption decreased by almost 4% compared to 2018 as a result of fuel oil displacement in the power sector due to the restart of several nuclear facilities, warmer than normal weather in the first months of 2019, and a consumption tax increase in October 2019 that put downward pressure on gasoline demand [26]. Japan consumed most oil in the transport (38%), industry

(24%), and non-energy use (16%) sectors in 2018. The share of the energy sector has declined from a peak of 19% in 2012 to 5% in 2018, as the sector started to replace oil with other fuels such as coal, natural gas, and nuclear power [27]. Gasoline, diesel, and naphtha accounted for the largest volumes of petroleum product demand in 2019 [28]. Crude oil imports declined by 13% on a year-on-year basis during the first seven months of 2020 [29]. The state of emergency imposed to curb the spread of COVID-19 in April and May slowed transport and significantly curbed gasoline and diesel sales. Sales of jet fuel remained depressed from September 2020 as many international flights remained grounded [30].

5.6.2 Natural Gas

In 2019, Japan's natural gas consumption reached an estimated 3.6 trillion cubic feet of natural gas per year (Tcf/year), about 4% higher than a decade ago (Fig. 5.5) [31]. Although natural gas consumption increased after the Fukushima catastrophe to replace lost nuclear capacity, natural gas consumption has declined since 2016 as several nuclear facilities have gradually returned to service and total electricity demand has declined [32]. The electricity sector was the largest consumer of natural gas, accounting for almost two-thirds of the market, followed by the industrial sector (20%) and the residential and commercial sector (13%) in 2019 [33]. EIA expects Japan's natural gas demand to decline further in 2020 as a result of the adverse effects of the COVID-19 pandemic, especially for the industrial and commercial sectors, and above-average temperatures during the winter of 2019–20 [34]. Over the next few years, the growth in demand for natural gas will depend on the pace of restarting nuclear facilities, the country's economic growth, and increasing competition from renewable energy sources.

5.6.3 Coal

Japan's coal demand was more than 181 Mt in 2019 (Fig. 5.6) [31]. Power generation, a major driver of coal use, accounted for 60% of coal consumption in Japan in 2018. The country's sizeable steel and iron ore industry accounted for the remainder of coal consumption [35].

Slow progress in restarting nuclear reactors and new coal-fired capacity supported coal imports and demand through 2019. However, the restart of several nuclear plants and the government's plan to replace older, inefficient coal plants with more efficient coal plants to meet its emissions target could be downside risks to coal demand in the coming years [36]. Coal is likely to remain a key part of Japan's power generation portfolio despite these limiting factors.

An additional 8 GW of new coal-fired generation is under construction and is expected to come online between 2020 and 2023. Some of this capacity is planned to replace older units [37]. Japan's coal demand is expected to decline in 2020 as a result of COVID-19 pandemic mitigation efforts and negative effects on energy and industrial demand.

5.7 Renewable Energy Prospect in Japan

Although the Japanese economy is very energy efficient compared to other OECD members, it is still one of the world's largest emitters of CO_2, both in absolute terms and per capita or per unit of Gross Domestic Product (GDP) [38] (Fig. 5.9). The energy sector is the largest emitter of CO_2, accounting for 49% of total emissions from fuel combustion in 2018, followed by transport (19%), industry (18%), services (7%), residential (5%), and other energy (3%) [39] (Fig. 5.10).

By fuel, the majority of energy-related CO_2 emissions came from coal combustion (41% of total emissions) and oil (35%), followed by natural gas (22%) in 2019 (Fig. 5.11).

Factors restricting the growth of renewable energy in Japan have mainly been [40]:

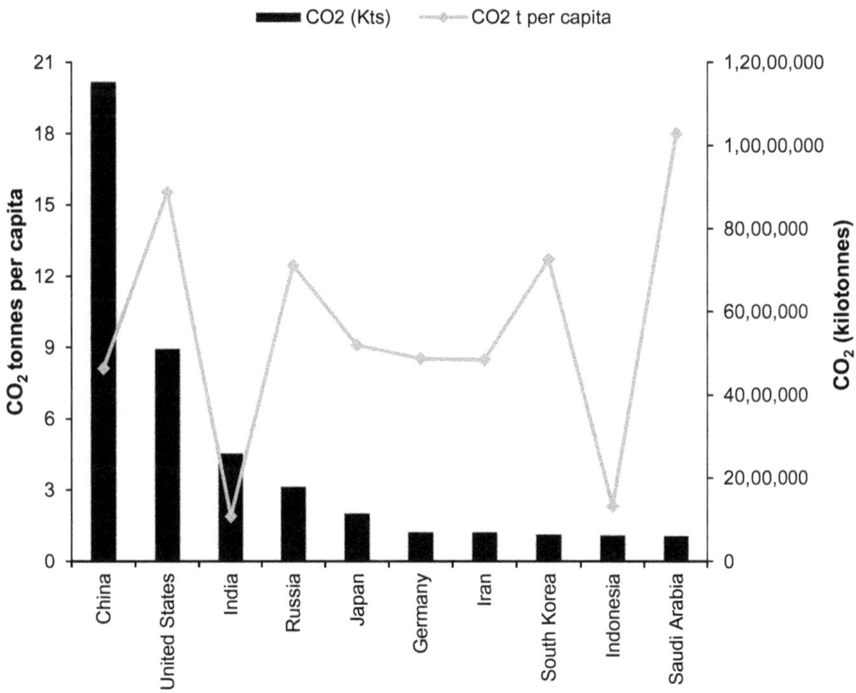

Fig. 5.9 CO_2 ranking of emissions and CO_2 per capita emissions (2019). *Source* prepared and adapted by the author on the basis of https://datosmacro.expansion.com/

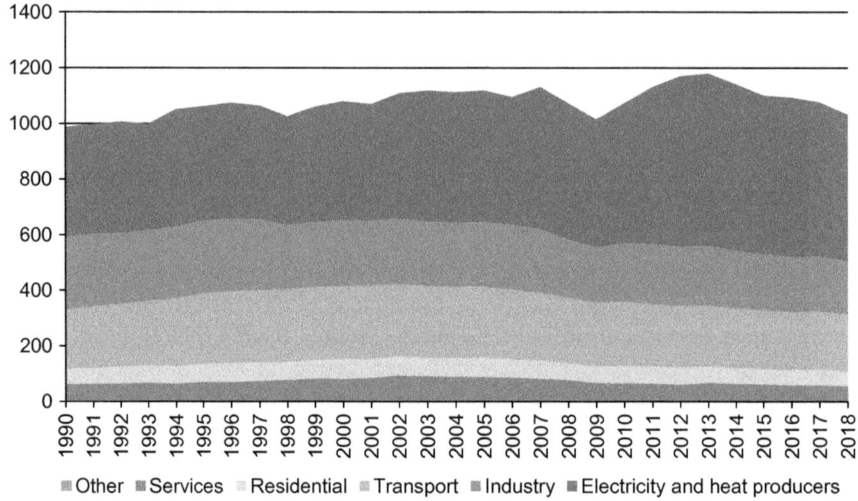

Fig. 5.10 CO_2 emissions by sector, Japan 2000–2018 *Source* prepared and adapted by the author on the basis of www.iea.org/statistics

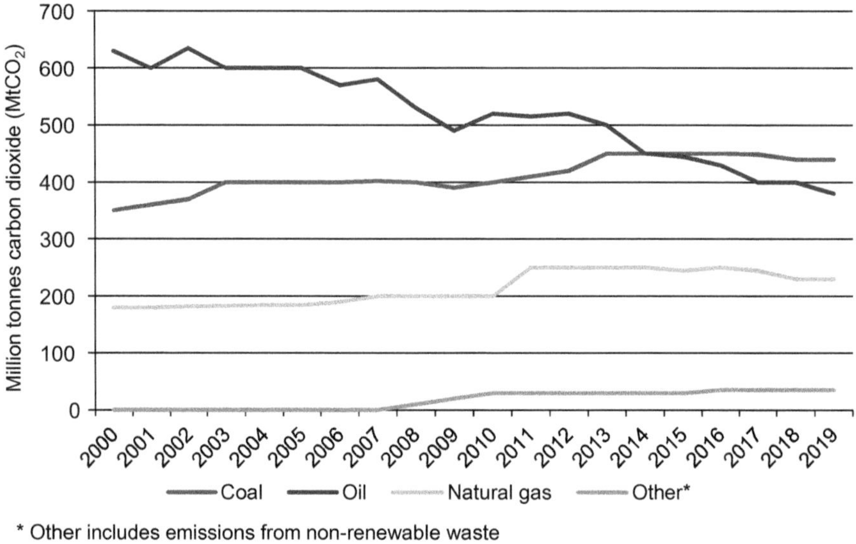

Fig. 5.11 Japan's energy-related CO_2 emissions by energy source, 2000–2019. *Source* prepared and adapted by the author on the basis of www.iea.org/statistics

- High cost of power generation. According to the International Renewable Energy Agency's Global Renewable Energy Database (2019) [41], The global levelised costs of electricity (LCOEs) obtained from bioenergy, geothermal, hydro, onshore, and offshore wind have fallen to the cost range of fossil fuels in 2010, and solar photovoltaic (PV) costs reached the cost range of fossil fuels in 2014. In addition, the absolute installed cost of commercial-scale solar PV remains twice as high as other countries' greenhouse gas emissions [42].
- Grid connection restrictions and site restrictions. In particular, the Japanese power grid has a longitudinal structure with few deviations and different frequencies in the east (50 Hz) and west (60 Hz). These conditions reduce the flexibility of the power systems. The International Energy Agency (2011) [43] analysed the penetration potentials of renewables in electricity demand and showed that Japan has the lowest penetration potential (19%) among the eight selected energy areas due to its low grid resilience, low co-ordination of the use of flexible resources and the expected use of renewable energy at the time of the analysis.

The extended FiT scheme in 2012 has been the main policy instrument to promote renewables. The result was an annual renewable capacity growth rate of 22% in 2017 since it was introduced [44]. Between 2010 and 2019, renewable capacity increased 2.7-fold (Fig. 5.12) and the share of renewables in the energy mix increased from 2.5% in 2010 to 9.1% in 2018. Strong growth in solar PV generation, from 0.3 to 6% of the generation mix, led to this increase. In fact, solar projects have accounted for around 90% of the capacity added since 2012, ranking Japan second in the world for solar capacity and third for solar generation [45–47], followed by bioenergy. Since the

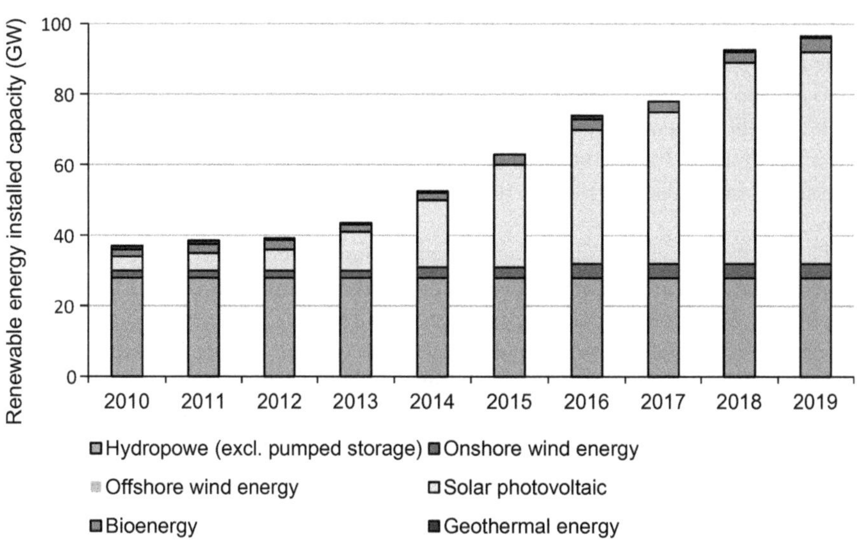

Fig. 5.12 Renewable installed capacity (GW), 2010–2019. *Source* prepared and adapted by the author on the basis of

introduction of the FiT scheme until March 2020, more than 70% of the cumulative purchased electricity comes from solar PV (275 TWh), followed by biomass (58 TWh, 14.8%), wind (44 TWh, 11.3%), small hydro (14 TWh, 3.6%) and geothermal (0.9 TWh, 0.2%) [48].

5.8 Framework of China's Renewable Energy Law and Policy

The deployment of renewable energy in Japan has been driven by plans to diversify electricity, achieve environmental and socio-economic goals, and increase energy security.

Events at the Fukushima nuclear power plant following the 2011 earthquake and tsunami marked an important turning point for Japan's energy policy [49] and diversification of the electricity mix.

The Rational Use of Energy Law (Energy Efficiency Law), established in 1979 and last updated in 2018, is the cornerstone of Japan's energy efficiency policies. It sets energy performance standards as well as reporting and energy management requirements for all sectors.

Japan's overall energy policy is set out in Strategic Energy Plans (SEPs), which were adopted under the 2002 Basic Act on Energy Policy. This energy policy is guided by the principles of energy security, economic efficiency, environmental sustainability, and safety.

The first energy policy review, the 4th Strategic Plan in 2014, aimed to minimise reliance on nuclear power and accelerate the adoption of renewable energy. At the same time, Japan pledged to reduce total CO_2 emissions by 26% between 2013 and 2030 [50] as part of its commitments to the Paris Agreement. Japan is currently on track to meet this emissions target: emissions from the power sector in 2019 were 3% lower than in 2010. But the power sector still accounts for about 40% of total emissions, and emissions peaked in 2013 when fossil fuels were replacing nuclear power.

The 2018 revision, the 5th Strategic Energy Plan, aimed to encourage joint energy efficiency improvements among businesses and the definition of carriers was revised (mainly to better encompass e-commerce retailers) to further encourage energy savings in freight transport. Its objectives are to achieve a more diversified energy mix by 2030, with a greater share for renewables and the resumption of nuclear power. It also aims to improve the efficiency of fossil fuel use and reduce energy demand [51].

The general assumptions of the 5th Strategic Energy Plans are [38]:

- Pursue the adoption of renewable energy resources the share of renewable energy in Total Primary Energy Supply (TPES) is expected to reach 13% in 2030, up from 8% in 2019. Renewable energy generation is expected to reach 24% in 2030, up from 19% in 2019.

- Minimise reliance on nuclear energy and work constantly to develop and secure new energy technologies, in view of the lack of domestic fossil fuel resources.

Since the introduction of Japan's 2030 target, the share of renewables in the TPES has increased from around 5% in 2014 to 8% in 2019. This growth was mainly driven by the expansion of renewables—mainly solar—in the electricity sector, thanks to a feed-in tariff (FIT) system introduced in 2012. From 2012 to 2018, renewable electricity capacity almost doubled, with solar PV accounting for 94% of this expansion. In 2019, renewables accounted for 19% of total electricity generation, up from 10% in 2012.

The energy efficiency of buildings has been regulated by the Energy Efficiency of Buildings Act since 2015.

On 25 December 2020, the Green Growth Strategy towards 2050 Carbon Neutrality was created, which sets out challenges and actions, formulates action plans and policies covering, among other taxes, regulatory reforms and standardisation, and international collaboration [52].

To accelerate the transition to clean energy, in July 2020 the government announced its intention to phase out its inefficient coal-fired power plants by 2030 [53, 54].

In 2009, the first FiT was introduced to remunerate excess solar production fed into the grid. This programme was expanded in 2012 (the "FiT Law") to cover all solar generation - not just excess - as well as hydro projects of less than 30 MW, wind, geothermal, biomass, and any other source recognised as renewable by the Japanese cabinet. The FiT's goal is to achieve 20–35% of energy supply from renewable sources by 2030 [55].

To control the rising FiT rate, an auction system was put in place to decide the purchase prices for large-scale solar PV (from 2017) and biomass power generation (from 2018). In 2020, the decision process for the purchase of offshore wind power will switch to an auction system.

The FiT system offers two solar PV options, with different durations and fixed prices. For a residential PV system (less than 10 kW), the FiT guarantees the purchase of excess energy at 42 Japanese yen (JPY) per kWh (including taxes) for 10 years. For non-residential PV systems (more than 10 kW), the FiT guarantees the purchase of all energy production at 40 JPY per kWh (including taxes) for 20 years [56, 57]. The FIT is financed by a tax on renewable energies levied on electricity consumers [58]. Table 5.1 shows the FiT purchase prices from 2012 to 2020 [59]. In 2017, the FIT introduced a tendering system for PV plants above 2,000 kW, and the fixed price is determined by the tendering process. In 2019, the PV plant size for tendering purposes was changed to more than 449 kW.

Table 5.1 The FiT Act purchase prices from 2012 to 2020

	PP < 10 kW	10 kW ≤ PP ≤ 1999		PP ≥ 2000 kW
Contract period	10 years	20 years		
Year	Tariff (JPY/kWh)			
2012	42	40		40
2013	38	36		36
2014	37	2		32
2015	33	29		29
2016	31	24		24
2017	28	21		Determined via the tendering process
2018	26	18		Determined via the tendering process
2019	24	10 kW ≤ PP ≤ 499 kW		PP ≥ 500 kW
		14		Determined via the tendering process
		10 kW ≤ PP ≤ 49	50 kW ≤ PP ≤ 249	PP ≥ 250 kW
2020	21	13	12	PP ≥ 500 kW

Note PP = PV plant capacity

5.9 Wind Turbine Installations

The deployment of wind energy has been considerably delayed due to grid connection restrictions, site constraints, time-consuming environmental assessment procedures, and the high costs involved [60]. As a result, offshore wind has been adopted as a particular solution of choice for large-scale renewable capacity. Although Japan does not have an offshore wind target for 2030, however, the foundations for the deployment of this energy source are being developed. In 2012, the feed-in tariff (FiT) for wind power was approved. Despite this, few FiT projects were installed offshore, due to market uncertainty and the complex Environment Impact Assessment (EIA) process. In 2017 and 2018, a number of changes were implemented with the aim of streamlining regulations and encouraging the development of offshore wind energy. In 2019, the Law on Promotion of the Utilisation of Marine Areas for the Development of Offshore Renewable Energy Generation Facilities came into force, followed by the establishment of the Public–Private Council for the Improvement of Industrial Competitiveness for Offshore Wind Energy Generation in July 2020 [61].

For all the above reasons, and with 4206 MW of wind installations by the end of 2020, including 65 MW of offshore wind power, wind power is becoming a source of support for Japan to reach the zero target and decarbonise its heavy industry, such as steel manufacturing and shipping (Fig. 5.13).

To date, floating offshore wind energy in Japan has mainly consisted of demonstration projects in Goto City in Nagasaki Prefecture (MOE project, 2 MW), in

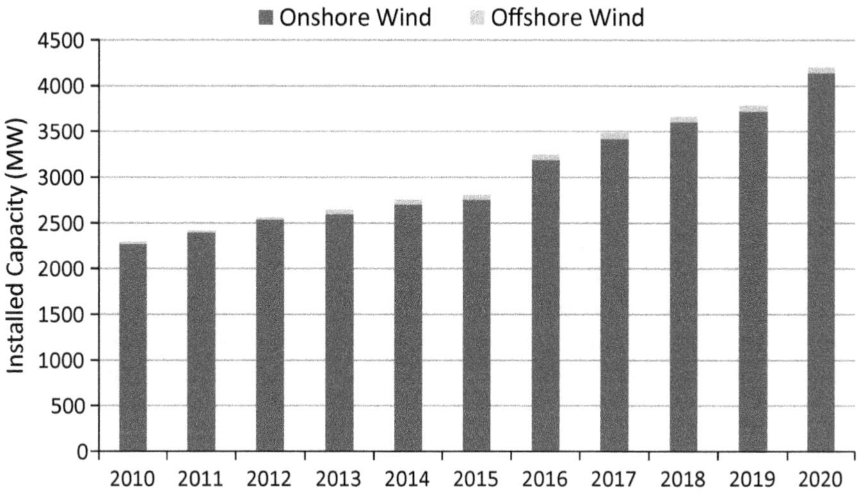

Fig. 5.13 Wind installed capacity in Japan from 2010 to 2020. *Source* prepared and adapted by the author on the basis of www.irena.org/solar

Fukushima Prefecture (METI project, 2 MW, 5 MW, 7 MW) [62], and in Kitakyushu City, Fukuoka Prefecture (NEDO project, 3 MW).

The growth trend is expected to continue beyond 2022 as more FiT projects are completed, causing Japan's total wind capacity to double by 2025. In addition, the government is expected to award 1.5 GW through an offshore wind tender next year, supporting capacity growth beyond 2022.

Currently, there are only a few offshore wind projects in operation in Japan. The first one was commissioned in 2003. Table 5.2 provides an overview of all projects [63]:

In total, eight projects have been installed. No project exceeds 16 MW and therefore the projects are all small compared to their European counterparts. In addition,

Table 5.2 Overview of Japanese offshore wind projects (2016)

Prefecture	Installed Capacity (MW)
Hokkaido	1.2
Yamagata	10
Fukushima	14
Ibaraki	14
Ibaraki	16
Chiba	2.4
Fukoaka	2.0
Nagasaki	2.0
Total	61.6

many of the projects are installed in port areas rather than in the open sea and many of them are merely pilot projects. For example, the Fukushima project represents one of the first applications of floating technology worldwide.

5.10 Solar PV Installations

Until the early 2000s, solar power upgrading was one of the country's top priorities. The implementation of the FiT has accelerated the rapid deployment of solar PV in Japan, especially for non-residential plants. Cumulative PV installation capacity doubled from 6.43 GW in 2012 to 12.11 GW in 2013, and the overall trend has been one of continued significant growth in the following years. The high fixed purchase price of FiT has attracted a large number of private investors and large companies to PV plants. Figure 5.14 shows Japan's PV installation capacity from 2012 to 2020. After 2015, Japan's annual PV installations started to show a downward trend. This can be attributed to the reduction of fixed prices. By the end of 2020, Japan's total solar PV capacity was 67 GW, ranking fourth in the world (Fig. 5.15), and solar PV accounted for 7.6% of total electricity consumption [64].

Prior to 2012, residential solar PV accounted for almost 85% of the total installed PV capacity. However, from this year onwards, the installed PV capacity is growing as a result of the increased number of non-residential PV plants, driven by the new FiT. In Fig. 5.16, it can be seen that the installation of non-residential PV plants decreases year by year, a consequence of the fact that the fixed price of the FIT for non-residential PV plants was reduced by 65% from 2012 to 2019 [65].

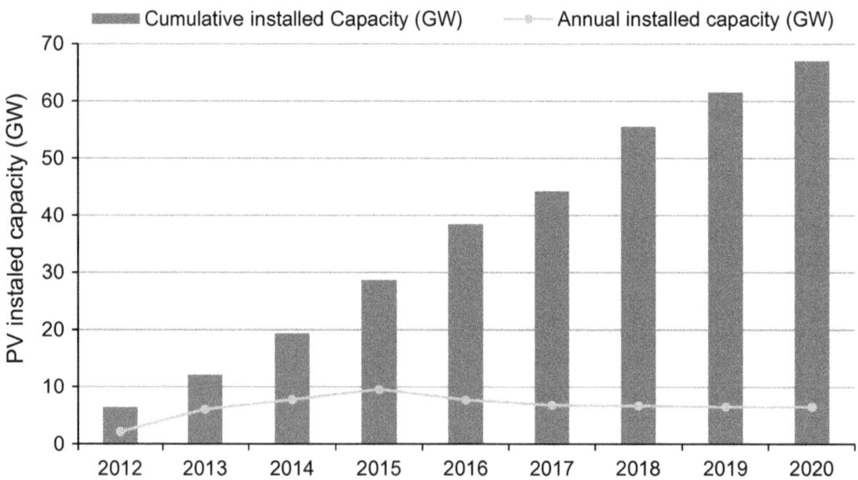

Fig. 5.14 Solar PV installed capacity in Japan from 2012 to 2020. *Source* prepared and adapted by the author on the basis of www.irena.org/solar

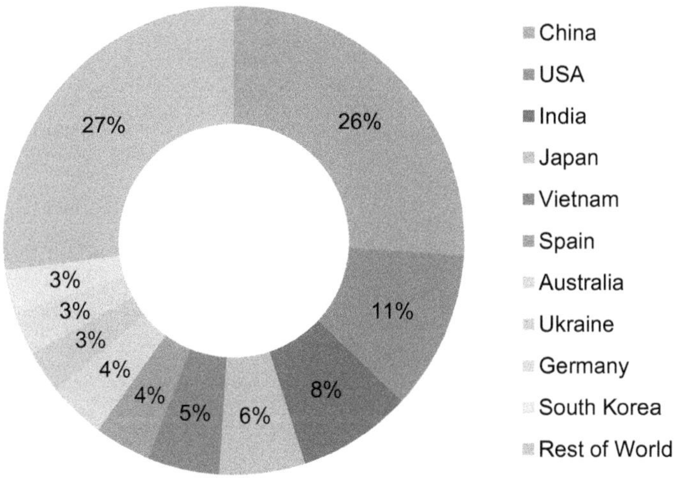

Fig. 5.15 Top 10 solar PV markets 2019. *Source* prepared and adapted by the author on the basis of www.iea.org/statistics

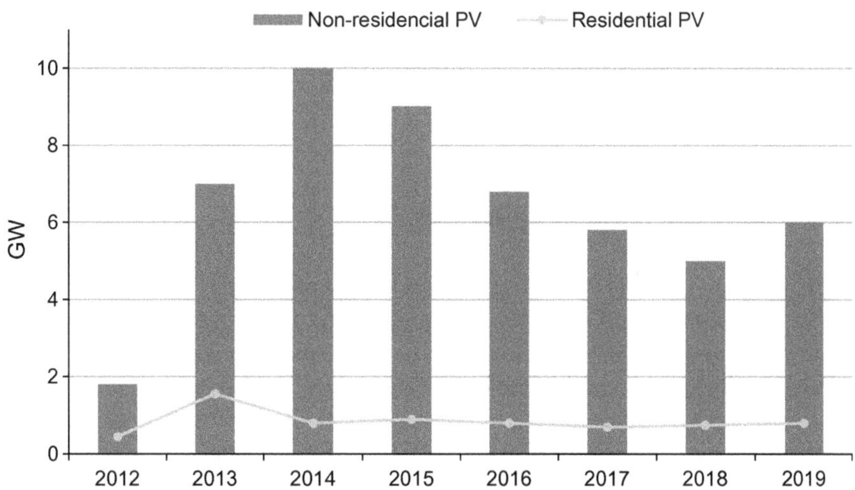

Fig. 5.16 Operational PV plants under the FiT from 2012 to 2019

On the other hand, Japan leads the world in terms of installed capacity and number of floating solar PV systems, due to the high FiT and lack of suitable land, followed by China, South Korea, and the UK. The country's many inland lakes and reservoirs are now home to 73 of the world's 100 largest floating solar plants, with a capacity of 123 MW [66]. Hyogo Prefecture in southern Honshu has almost 40,000 lakes and already hosts almost half of the floating solar capacity of the world's 100 largest plants. Ciel & Terre is the largest installer in the market, with more than 50% of

Table 5.3 The world's top 10 floating solar plants

Location	Scale (kW)	Country
Coal mining subsidence area of Huainan City	40,000	China
Coal mining subsidence area of Huainan City	20,000	China
Yakamura solar power plant	13,700	Japan
Pei County	9982	China
Umenoki	7550	Japan
Hirotani Ike Floating Solar Plant	6800	Japan
Jining GCL	6776	China
Queen Elisabeth II Reservoir	6338	UK
Cheongpung Lake	3000	South Korea
Otae Province	3000	South Korea

the total number of plants worldwide, and 38 of the 54 largest plants in Japan. Its projects are mainly located in the western part of Japan because it can utilise many reservoirs in that area and take advantage of good grid connections, compared to the north of the country where snowfall is frequent [67]

Japan's largest floating solar power plant is located behind the Yamakura Dam in Ichihara (Chiba Prefecture). It occupies 18 hectares, can supply power to 5,000 households, and saves more than 8000 tonnes of CO_2 per year (Table 5.3) [64].

5.11 Solar Thermal Installations

In 2011, Japan had 3.35 GWt of Solar Water Heating (SWH) capacity in operation [68]. However, most of these systems were installed in the 1970s–80s and the number of new installations has been steadily decreasing. In the framework of the national programme "Sunshine Project", solar heating and cooling technology and solar thermal power generation have been promoted since 1974 [69]. The share of solar thermal in domestic hot water supply in Japanese residential buildings increased from almost zero in 1970 to around 8% in 1980, while its share is less than 2% in 2015.

Figure 5.17 shows the number of installed solar water heating units and solar thermal systems in Japan for 1975–2017 [70].

Although the Japanese solar heating market experienced a boom in the late 1970s and early 1980s, the number of systems declined after that period. With the rising price of imported oil in the mid-1980s, solar thermal heating units lost their level of competitiveness against conventional heating systems. At the end of the period 2005–2012, the solar thermal market was more or less stagnant, due to limited government support, relatively high prices, and the financial crisis that hit Japan in the second half of 2009. Only 2008 saw a slightly higher installation.

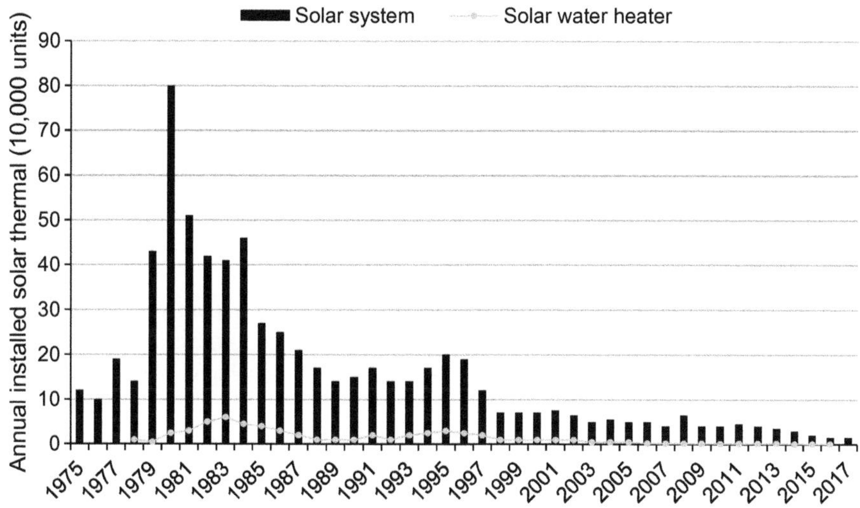

Fig. 5.17 Installed solar heating units (10,000 units) in Japan, 1975–2017

In the 2010s, and especially after the Fukushima nuclear disaster in 2011, Japan faced an energy challenge. Due to the rapidly rising price of crude oil and also people's focus on sustainable energy, solar thermal experts expected the solar thermal market to skyrocket after Fukushima. But it turned out to be a boost for solar power generation, caused by the feed-in tariff [71].

Currently, the total number of installations in Japan amounts to about 4.2 million units, of which about 25% are installed in the Kyushu region of southern Japan.

5.12 Conclusions

In October 2020, Japan's new Prime Minister declared that by 2050 Japan will aim to reduce greenhouse gas emissions to zero and achieve a carbon–neutral, decarbonised society.

Over the past decade, Japan has made considerable progress in implementing its vision of an efficient, resilient and sustainable energy system, basing its energy policy on the principles of energy security, economic efficiency, environmental sustainability, and safety.

The gradual restart of nuclear power generation, expansion of renewable energy, and increased energy efficiency have reduced the need for fossil fuel imports and contributed to a steady decline in greenhouse gas (GHG) emissions. These reached an all-time high in 2013, as fossil fuels filled the gap from nuclear power plants after the Fukushima accident. By 2018, GHG emissions had decreased by 12% compared to 2013, returning to the same level they were at in 2009.

Despite all efforts, Japan remains heavily dependent on imported fossil fuels. In 2019, fossil fuels accounted for 88% of total primary energy supply.

To achieve the goal of carbon neutrality by 2050, Japan will need to substantially accelerate the deployment of low-carbon technologies, address regulatory and institutional barriers, and improve competition in its energy markets. In the energy sector, the main challenge is to address grid constraints, including through better connectivity between Japan's regional grids. Regulatory reform can help to improve the functioning of the electricity system, thus enabling the integration of a higher share of variable renewable energy.

Japan has experienced a rapid expansion of solar PV in recent years, driven by generous feed-in tariffs. However, more efforts are needed to develop other renewable technologies, such as wind and geothermal, for which Japan has large energy potential.

Wind and geothermal energy have great potential in Japan and can complement the emission reductions being achieved with the expansion of nuclear and solar power.

References

1. Ministry of Environment (2020) The long-term strategy under the Paris agreement. Last access 11 Apr 2020. https://www.env.go.jp/press/111913.pdf
2. Ministry of Economy (2018) Technology and industry, strategic energy plan (provisional translation). Last access 11 Apr 2020. https://www.enecho.meti.go.jp/en/category/others/basic_plan/5th/pdf/strategic_energy_plan.pdf
3. Obane H, Nagai Y, Asano K (2020) Assessing land use and potential conflict in solar and onshore wind energy in Japan. Renew Energy 160:842–851
4. Julia Máxima Uriarte. Japón. Caracteristicas.co. 10 de marzo de 2020. https://www.caracteristicas.co/japon/. Accessed on 16 May 2021
5. World Bank. Electric power consumption (kWh per capita). Available online: https://data.worldbank.org/indicator/EG.USE.ELEC.KH.PC?locations=JP-CN. Accessed on 14 May 2021
6. World Bank. GDP (current US$)-Japan. Available online: https://data.worldbank.org/indicator/NY.GDP.MKTP.CD?locations=JP&most_recent_value_desc=true. Accessed on 14 May 2021
7. World Bank (2020) GDP growth (annual %)—Japan. World Bank National Accounts Data (database), World Bank, Washington, DC. https://data.worldbank.org/indicator/NY.GDP.MKTP.KD.ZG?locations=JP. Accessed on 8 Apr 2021
8. Expansión. Datosmacro.com. Japón: El país con más deuda. Available online: https://datosmacro.expansion.com/analisis/los-peores/pais-mas-deuda#:~:text=Jap%C3%B3n%20es%20el%20pa%C3%ADs%20con,%2C6%25%20de%20su%20PIB. Accessed on 14 Apr 2021
9. International Energy Agency, Electricity Information 2020; International Energy Agency, Oil Market Report, August, 11, 2016, page 9; Federation of Electric Power Companies of Japan, Electricity Generated and Purchased (Bulletin), FY 2010
10. OECD (Organisation for Economic Co-operation and Development) (2020) OECD economic outlook: vol 2020, Issue 1, OECD Publishing, Paris. https://doi.org/10.1787/0d1d1e2e-en. Accessed on 14 Apr 2021
11. The World Bank, Source: Global Solar Atlas 2.0 (2019), Solar resource data: Solargis. https://solargis.com/maps-and-gis-data/download/japan. Accessed on 14 Apr 2021
12. IEA. Japan 2021. Energy Policy Review. https://www.iea.org/reports/japan-2021 [accessed on October 25, 2020].

13. International Energy Agency, Electricity Information 2020; FACTS Global Energy, "Japan's New Power Generation Mix to Remain Unchanged," April 3, 2018
14. EIA estimates derived from International Energy Agency Electricity Information 2020
15. Japan's Ministry of Economy, Trade, and Industry, Agency for Natural Resources and Energy, Japan's Strategic Energy Plan, April 12, 2018; Japan's Ministry of Economy, Trade, and Industry, 5th Strategic Energy Plan Outline. Accessed on 28 Aug 2020); World Nuclear News, "Japanese Cabinet approves new basic energy plan," July 3, 2018
16. BP Statistical Review of World Energy 2020; EIA estimates derived from International Energy Agency, Electricity Information 2020
17. BP Statistical Review of World Energy 2020; PV Magazine, "Japan clears up uncertainty over its renewable energy FIT," May 31, 2016; Japan Renewable Energy Foundation: A Statement of the Basic Energy Plan of Japan, April 21, 2014
18. PV Magazine, Japan may surpass 2030 PV target of 64 GW within two years—RTS, December 12, 2017
19. Oil & Gas Journal, Worldwide Look at Reserves and Production, December 2, 2019
20. International Energy Agency, Monthly Oil Database Service. Accessed on 25 Apr 2021
21. World Bank data: GDP growth [Accessed on 5 Sept 2020); International Monetary Fund, "2019 Article IV Consultation for Japan," January 14, 2020, page 1; S&P Global Platts, "Japan Q1 crude steel production volumes lowest in over 10 years, worse to come," April 28, 2020; Reuters, "Japan exports post worst fall in two years as trade war bites," January 22, 2019; Reuters, "Japan says GDP growth could slow to 0.9% on weakening global demand," July 29, 2019
22. U.S. Energy Information Administration, International Energy Statistics; Global Trade Tracker [accessed on September 5, 2020)
23. Global Trade Tracker. Accessed on 5 Sept 2020
24. U.S. Energy Information Administration, Short-Term Energy Outlook September 2020
25. International Energy Agency, Oil 2020, page 28; International Energy Agency, Energy Policies of IEA Countries: Japan 2016 Review, page 61; FACTS Global Energy, Asia Pacific Petroleum Databook 1: Supply and Demand, Spring 2020, pages 40–41 and Fall 2019, p 41
26. FACTS Global Energy, Asia Pacific Petroleum Databook 1: Supply and Demand, Spring 2020, pp 40–41 and Fall 2019, page 41; Reuters, "Japan eyes tax breaks for car buyers to offset impact of sales tax hike," September 6, 2018; U.S. News and World Report, "Japan Ups Sales Tax to 10% Amid Signs Economy Is Weakening," October 1, 2019
27. U.S. Energy Information Administration using International Energy Statistics, Oil Information 2020 data [Accessed on 13 Nov 2020]; International Energy Agency, Energy Policies of IEA Countries: Japan 2016 Review, page 61
28. FACTS Global Energy, Asia Pacific Petroleum Databook 1: Supply and Demand, Spring 2020, p 41
29. Global Trade Tracker José Pablo Martínez Garcia
30. S&P Global Platts, "JAPAN DATA: Crude imports hit 54-year low on pandemic, non-Mideast oil squeezed out," July 31, 2020
31. U.S. Energy Information Administration, International Energy Statistics
32. FACTS Global Energy, Global LNG Monthly Report, August 16, 2018, page 3; International Energy Agency, Gas 2018: Analysis and Forecasts to 2023, pp 28–30; Newsbase, AsianOil, "Japan's gas demand continues to ease," August 1, 2018, page 8; Newsbase, AsianOil, "Japan's LNG imports down, while prices rise," July 25, 2018, page 11
33. FACTS Global Energy, East of Suez Gas Databook: Japan Natural Gas Outlook, August 2020, p 10
34. Reuters, "Japan city gas demand falls 20% in May due to COVID-19," June 11, 2020; S&P Global Platts, "Analysis: Japan's LNG demand unlikely to recover fully from COVID-19 in H_2," June 12, 2020
35. International Energy Agency, Coal 2019: Analysis and Forecasts to 2024, page 90; Coal 2018: Analysis and Forecasts to 2023, p 23

36. Reuters, "Japan to accelerate closure of old coal power plants," July 2, 2020; Reuters, "Japan to shut or mothball 100 ageing coal-fired power plants: Yomiuri," July 1, 2020; International Energy Agency, Coal 2019: Analysis and Forecasts to *2024*, page 91; The Japan Times, "Japan's phaseout plan forces utilities to rethink reliance on coal," July 16, 2020
37. International Energy Agency, Coal 2019: Analysis and Forecasts to 2024, page 91; Japan Electric Power Information Center, Inc., The Electric Power Industry in Japan 2020, p 36
38. WRI (World Resources Institute) (2020) Climate Data Explorer (CAIT). http://cait.wri.org. Accessed on 27 Marc 2020
39. IEA (International Energy Agency) (2020) CO_2 emissions by product and flow. IEA CO_2 Emissions from Fuel Combustion Statistics (database), IEA, Paris. www.iea.org/statistics
40. Shiraki H, Sugiyama M, Matsuo Y, Komiyama R, Fujimori S, Kato E, Oshiro K, Silva DH (2021) The role of renewables in the Japanese power sector: implications from the EMF35 JMIP. Sustain Sci 16(2):375–392
41. International Renewable Energy Agency (2019) Renewable power generation costs in 2018
42. Kimura K, Zissler R (2016) Comparing prices and costs of solar PV in Japan and Germany, Tokyo
43. International Energy Agency (2011) Harnessing variable renewables: a guide to the balancing challenge. Organisation for Economic Co-operation and Development, Paris
44. METI (2019) Japan's energy 2018: 10 questions for understanding the current energy situation, Ministry of Economy, Trade and Industry. https://www.enecho.meti.go.jp/en/category/brochures/pdf/japan_energy_2018.pdf
45. IRENA (2020) Renewable energy statistics 2020, International Renewable Energy Agency, Abu Dhabi
46. METI (2017) FY2016 annual report on energy (Energy White Paper), Ministry of Economy, Trade and Industry. https://www.meti.go.jp/english/report/downloadfiles/2017_outline.pdf
47. IRENA (2019b), Renewable energy statistics 2019, International Renewable Energy Agency, Abu Dhabi.
48. Agency for Natural Resources and Energy (2020) Website for information disclosure on the Feed-in Tarrif (in Japanese). Agency Nat Resour Energy. https://www.fit-portal.go.jp/PublicInfoSummary. Accessed 8 July 2021
49. Kucharski JB, Unesaki H (2017) Japan's 2014 strategic energy plan: a planned energy system transition. J Energy
50. METI (2018) 5th strategic energy plan, Ministry of Economy, Trade and Industry. https://www.enecho.meti.go.jp/en/category/others/basic_plan/5th/pdf/strategic_energy_plan_outline.pdf
51. METI (2018) Strategic Energy Plan, METI, Tokyo. www.enecho.meti.go.jp/en/category/others/basic_plan/5th/pdf/strategic_energy_plan.pdf
52. METI (2020) Green Growth Strategy towards 2050 Carbon Neutrality" Formulated, METI, Tokyo. www.meti.go.jp/english/press/2020/1225_001.html
53. Reuters (2020) Japan to shut or mothball 100 ageing coal-fired power plants: Yomiuri. Reuters, www.reuters.com/article/us-japan-powerstation-coal/japan-to-shut-or-mothball-100-ageing-coal-fired-power-plants-yomiuri-idUSKBN243074.Accessed 9 July 2020
54. The Economist (2020) Japan promises to be carbon-neutral by 2050. The Economist. www.economist.com/asia/2020/10/29/japan-promises-to-be-carbon-neutral-by-2050.Accessed on 14 Apr 2021
55. Muhammad-Sukki F, Abu-Bakar SH, Munir AB, Mohd Yasin SH, RamirezIniguez R, McMeekin SG, Stewart BG, Sarmah N, Mallick TK, Abdul Rahim R, Karim ME, Ahmad S, Mat Tahar R (2014) Feed-in tariff for solar photovoltaic: the rise of Japan. Renew Energy 68:636–643. https://doi.org/10.1016/j.renene.2014.03.012
56. International Energy Agency (2019) Japan solar PV auctions. https://www.iea.org/policies/6563-japan-solar-pv-auctions?country=Japan&region=AsiaPacific. Accessed 7 May 2021
57. METI (2012) Feed-in Tariff Scheme in Japan
58. METI (2020) Outline of feed-in tariff act. https://www.enecho.meti.go.jp/category/saving_and_new/saiene/kaitori/surcharge.html

59. METI (2020) Feed-in tariff act; purchase price, period, etc. (FY2012-FY2020). https://www.enecho.meti.go.jp/category/saving_and_new/saiene/kaitori/kakaku.html#2019
60. Mizuno E (2014) Overview of wind energy policy and development in Japan. Renew Sustain Energy Rev 40:999–1018. https://doi.org/10.1016/j.rser.2014.07.184
61. GWEC—Global Wind Energy Council (2020) Global offshore wind report 2020. GWEC, Brussels, Belgium
62. Fukushima Offshore Wind Consortium, "Fukushima Floating Offshore Wind Farm Demonstration Project". http://www.fukushima-forward.jp/pdf/pamphlet.pdf
63. Li A, Xu Y (2018) The governance for offshore wind in Japan. Elsevier, ScienceDirect: Amsterdam, The Netherlands, ISBN 1876-6102
64. IRENA (2020) Solar energy data [WWW Document]. https://www.irena.org/solar
65. Wen D, Gao W, Kuroki S, Gu Q, Ren J (2021) The effects of the new Feed-In Tariff Act for solar photovoltaic (PV) energy in the wake of the Fukushima accident in Japan. Energy Policy 156:112414
66. Solarplaza. https://solarassetmanagement.asia/top-70-floating-solar-plants-2018
67. Japan Today, April 18, 2017. https://JapanToday.com/category/features/lifestyle/Japan-sees-potentialin-solar-power
68. REN21 (2013) Renewables 2013 global status Report - renewable energy policy network for the 21st Century. REN21 http://www.ren21.net/portals/0/documents/resources/gsr/2013/gsr2013_lowres.pdf
69. Noguchi T (1985) Overview on thermal application of solar energy in Japan. Sol Wind Technol 2(3/4):155–171
70. Solar System Development Association of Japan (SSDA), online data. http://www.ssda.or.jp/
71. Halawa E, Chang KC, Yoshinaga M (2015) Thermal performance evaluation of solar water heating systems in Australia, Taiwan and Japan—a comprehensive review. Renew Energy 83:1279–1286

Chapter 6
Wind Energy and Solar PV Developments in the USA

6.1 Introduction

The United States of America (USA) is an independent country, one of the four largest in terms of land area and the third largest in terms of population. Comprising 50 states and a federal district, the United States extends across central and extreme northwestern North America, the Hawaiian archipelago, and some territories in the Caribbean and Oceania.

The US is considered the richest, most powerful, and influential country on Earth. Its nominal GDP represented about 1/4 of the world's nominal GDP in 2019. In historical terms, the United States is a relatively young superpower, with only 70 years separating its coronation from today.

The US is the world's largest producer of oil and the 14th largest by reserves; the third largest producer of coal and the largest by reserves, with about 27% of total production, and, the largest producer of natural gas, where it is the third largest by reserves.

Both energy production and consumption reached record-high values in 2018. In 2019 energy production continued to grow by around 6% while the consumption decreased by about 1%, this was the first time the production exceeded the consumption on an annual basis since 1957.

Globally, the United States is the largest energy consumer, with approximately 17% of the world's total, in line with its share of world GDP. Its emissions have been on a downward trend, albeit with ups and downs, since 2007. In absolute terms, it is the world's second largest emitter of greenhouse gases (GHG) after China.

In 2019, the US annual energy consumption from renewable energy sources was higher than the coal consumption for the first time since 1885. The addition of new wind and solar capacity in unprecedented quantities in the United States, together with the potential of hydroelectric and nuclear power, accounted for 40% of total electricity production in the United States in 2020.

Currently, there is over 105 GW of wind power installed capacity in the United States; this makes the US the second largest country in the world in installed capacity.

T. Muneer et al., *The Coming of Age of Solar and Wind Power*, Green Energy and Technology, https://doi.org/10.1007/978-3-030-92010-4_6

As the largest source of renewable electricity, wind power also provides significant environmental benefits. Wind power reduced the US carbon dioxide emissions by 198 million metric tonnes in 2019.

The US solar market is growing rapidly as the industry expands innovative new technologies and reduces manufacturing and installation costs. Public interest in clean technologies and renewable energy has increased the demand for solar power. The movement towards a sustainable energy infrastructure has encouraged consumers and businesses to demand renewable energy sources, such as solar power.

6.2 Geographical Background

The geography of the United States is mainly determined by the extent of its relief. The USA is the second largest country in America (after Canada) and the fourth largest in the world. It covers an area of 9,828,675 km^2, of which 469,496 km^2 is water and has a coastline of 19,924 km [1].

The USA is bordered to the north by Canada, to the south by Mexico, to the east by the Atlantic Ocean, and to the west by the Pacific Ocean. It is made up of 50 states, including Hawaii, which is a state that is not on the mainland but is located on a Polynesian archipelago in the Pacific [1].

The relief of the USA is very varied, its main characteristic being that it is located on the North American plate, which changes its relief from west to east. In the west, its coastline borders the Pacific plate, making it a mountainous area that is also characterised by the frequent presence of volcanic phenomena and earthquakes. Thus, as it moves away from the Pacific plate, the terrain becomes more stable, with large plains in the centre and plateaus in the east of the country.

The relief of the USA is made up of a series of mountain ranges at least 2000 km long, known as the Pacific Coast Ranges. To the east are the high-altitude Appalachian Mountains, and to the west are the Rocky Mountains, the Sierra Nevada, and the Cascade Range. Here you have the combined heights of the 3250 m Baker Mountains, the 4390 m Rainier Mountains, and the 1300 to 1500 m high plateau of the Great Basin. There are also the Great Plains which are about 2500 m wide [2].

The hydrography of the USA is centred on three main river basins: Mississippi, Missouri, and Ohio, which cover an area of 2,980,000 km^2. Most of its rivers originate in the mountains and flow into the sea, although very few originate in the interior of the Rocky Mountain system [3].

The U.S. river network is made up of short but fast-flowing rivers. The Mississippi is an exception, with a catchment area of 3,328,000 km^2 and a length of 3778 km, it is a fully navigable river. Its tributaries are Missouri, Ohio, and Arkansas. Other major rivers in the United States include the Colorado River, the Rio Grande, the Yukon River, the Columbia River, the Hudson River, the Tennessee River, the St. Lawrence River, the Arkansas River, the Savannah River, the Illinois River, and the Potomac River, among others.

The Atlantic coasts are quite high, with inlets stretching all the way to Cape Hateras that turn into beautiful fine sand dunes. The Florida Peninsula has a coastline of coral islets and the Gulf of Mexico coast is very marshy and features the wide Mississippi River Delta [4, 5].

Given the large geographic area of the United States, different types of climates can be found in different parts of the country, ranging from tropical to cold and dry. In general, the southern region is the warmest and the northern region the coldest. The main types of climates existing in the country are:

Humid continental climate: This mainly covers the north of the country. Winters are usually mild, but temperatures often drop sharply due to the arrival of polar air, producing frequent snowfalls on the highest peaks. Precipitation usually occurs in the summer, which is warm and humid.

Dry climate: Affects the central United States, from east of the Mississippi River to near the Rocky Mountains.

Oceanic climate: The western US, from southern Alaska to California, has a temperate climate with high moisture content, resulting in winters with abundant rainfall.

Mediterranean climate: Typical of central and southern California. This climate is similar to that of the Mediterranean coast of Europe, with mild temperatures. This microclimate is due to the influence of the Pacific Ocean and the proximity of the Rocky Mountains.

Humid subtropical climate: Covers from the southern half of the west coast to Mexico, the Gulf of Mexico, and Florida, including some inland plains. Temperatures are warm year-round and humidity is very high. In southern Florida, the climate can be considered monsoonal with frequent hurricanes.

Dry subtropical climate: The areas with this climate are Texas, New Mexico, Nevada, and Arizona. Large deserts abound here.

Tropical climate: Climate typical of Hawaii. It is characterised by mild and constant temperatures of around 24–25 °C throughout the year. Rainfall is generous, which favours a very peculiar type of flora and fauna.

Climate of Alaska: The state of Alaska has a variety of climates. In the coastal strip, it is temperate. But in the rest of the territory, it is subarctic, with low temperatures and high humidity, the minimum in winter easily reaches −30 °C which causes areas with abundant snow and ice [6–8].

6.3 Historical Background

The United States (USA) has not very old but intense history, with a mixture of cultures from different parts of the world.

At the end of the nineteenth century, when the United States was already a regional power in America, it also became one of the world´s major economic power. Its domestic market, connected by railroad networks and telegraph cables, was self-sufficient thanks to its vast natural resources and arable land, as well as its numerous ports on the Atlantic and Pacific Oceans. In addition to this economic self-sufficiency, the location of the United States between two oceans and its stability as a region, help to maintain the country far from any enemy. None of this gave the US any reason to become embroiled in global dynamics, so the Americans maintained limited military and foreign policy power.

However, the isolation that had allowed the United States to prosper was broken in 1898, when they assisted the Cuban resistance in its struggle for independence from Spain and went to war. The easy victory against Spain marked the birth of the United States as a world power and established it as a regional power in the Americas [9–11].

On the eve of World War I, America's economic and industrial capacity exceeded that of Europe combined, however, this economic supremacy was not an international power yet. This situation changed with World War I, which reduced Europe's power, devastating it psychologically, economically, and demographically. By contrast, the United States developed economically by trading with the European powers in war.

This was possible thanks to the neutral position the US had during the first years of the war. But, in 1917 the United States entered the war on the Allied side due to Germany´s attacks. This fact broke the tradition of isolationism and neutrality towards Europe.

When the war came to an end, the former president of the US Wilson, made a number of proposals to ensure lasting peace in the future. These proposals were known as the "Fourteen Points" and were focused on the self-determination of the people, democracy, and international cooperation, and also had the objective to establish a new international order based on free trade, freedom of navigation, and disarmament. These proposals have been the foundations of liberal internationalism, which has characterised US foreign policy ever since.

The Fourteen Points established a new international organisation, the League of Nations, whose purpose was to encourage the resolution of international conflicts through dialogue. But despite pushing for its creation, the United States decided not to join the League of Nations, fearing a loss of sovereignty and wanting to maintain its international neutrality [12].

Although the United States decided not to join the League of Nations, it led to the international order linking the rest of the world with economic and commercial ties. In 1929 the US stock market collapsed, causing a global financial crisis, known as the Great Depression. This crisis was caused precisely by the mentioned economic connections. As the international order of the League of Nations broke down, US President Roosevelt decided to remain strictly neutral and to priories the domestic challenges of the financial crisis. This changed in 1940 when the US economy was established and World War II broke out in Europe, with the shift from neutrality to non-belligerency to provide help to countries at war against Germany and Italy [13–15].

However, the non-belligerency ended brusquely on 7 December 1941, when Japan suddenly attacked the US naval base at Pearl Harbor in Hawaii. The US entered the war as an ally of the UK, the USSR, and China. It was the only power fighting on two separate fronts at the same time. There were over 16 million Americans directly or indirectly involved in the war and producing the goods that sustained the allies [14–17].

The war finished in 1945 when the US launched atomic bombs in the cities of Hiroshima and Nagasaki in Japan. The United Stated already knew that it was not possible to maintain the country isolated in a world with fighter planes and nuclear weapons, therefore they decided that they had to lead the new international order to ensure the security of the country. The first step was to create the successor of the League of Nations, the United Nations (UN) in 1945 [17–19].

The United States emerged from the war practically unscathed and more prosperous than ever, having not only the monopoly on nuclear weapons but also the most advanced military and commercial technology. Only the Soviet Union, which dominated Eastern Europe after its victory against Germany and whose communist ideology had worldwide support, was able to challenge the US leadership. The US and the Soviet Union gained the new status of superpowers, even though their values were virtually opposed, with the liberal internationalism and capitalist economics of the US and the communism of the Soviet Union. Thus, the Cold War started a period in which the two superpowers vied for global influence [10–16].

The détente did not prevent the US from entering the Vietnam War in 1965 to prevent the communist faction from seizing power in the country. After more than a decade of intervention, billions of dollars spent, and nearly 60,000 American casualties, the US failed to achieve its goal. The tactical complexity, US civil society's opposition to the war, and the prolongation of the conflict until 1973, left lasting scars on US foreign policy [20, 21].

When the USSR unexpectedly collapsed in 1991, the United States was left as the sole hegemonic power and began to build a world in its own image. In the 1990s, as democracy and economic liberalism extended in the world, the US reinforced its alliances, with many former Eastern bloc countries joining NATO. This hegemony led the US to intervene even in conflicts where its interests were not at stake, as in the Yugoslav wars [22, 23].

At the same time, the US augmented its presence in the Middle East, a region of strategic importance because of its large oil reserves. Moreover, this region is home to Israel, one of the US's main allies, which was the first country to recognise the New Hebrew state in 1948. The relationship between the two is based primarily on historical and cultural ties, as well as mutual security interests.

When Kuwait was invaded by the Iraqi regime of Saddam Hussein in 1990, an international coalition led by the US defeated Iraq in the Gulf War. The US kept its troops in Iraq causing mistrust in some local groups. Al Qaeda, a jihadist group led by Osama Bin Laden, emerged during this period and would become the world´s largest terrorist organisation with the attack on the Twin Towers in New York in 2001. As a consequence of that attack, the president of the US George W. Bush started the "War of Terror", a campaign against not only terrorist groups but also countries suspected of collaborating with them.

After the attack of the Twin Towers, President George W. Bush started the "War of Terror", in order to crack down on counter-terrorist groups such as Al Qaeda, as well as with countries that collaborate with these groups. During this period, the US adopted a more interventionist and unilateralist foreign policy.

Afghanistan, where Al Qaeda was headquartered, was the first theatre of this campaign. The United States invaded Afghanistan when the Taliban refused to hand over hand over Bin Laden. Even though the US soon controlled the capital of Afghanistan Kabul, it was very difficult to maintain control over the country, forcing the US to extend their presence in the country. Afghanistan is currently the longest war in which the US has been involved in its history [24, 25].

The second theatre of the war on terror was Iraq, where members of the Bush administration mistakenly believed there were weapons of mass destruction. Despite having no evidence and virtually no international support, the US invaded Iraq in 2003 in violation of international law. This invasion had a similar outcome to that of Afghanistan: despite quickly taking the capital and toppling Hussein, the difficulty of maintaining control of the country would prolong the war. When the last US troops left Iraq in 2011, they left behind a country more violent and unstable than a decade earlier, the breeding ground for terrorist organisations such as Da'esh to emerge.

Approximately two decades after the end of the Cold War, US supremacy, whose leadership was already in question after the invasion of Iraq, received a major setback in the economic crisis of 2008.

In 2008, US citizens elected Barack Obama as their new president, the first African-American in office. In his tenure as president, Obama focused on ameliorating the worst economic recession the American people have known since the Great Depression of the 1930s.

As Western economies collapsed and US leadership crumbled, other powers such as China emerged. President Obama sought to move away from the Bush legacy by adopting a more multilateral stance, relying on NATO to intervene in the Libyan war in 2011, for example. Obama also sought to improve diplomatic relations with historic US enemies such as Cuba and Iran, with whom he negotiated a deal to limit its nuclear programme in 2015 [26–28].

Donald Trump launched his candidacy for the US presidency in June 2015 with the famous "Make America Great Again", a slogan that became the centrepiece of all his speeches. Finally, on 20 January 2017, Donald Trump became the 45th President of the United States.

Donald Trump's four years in office had a great impact on the climate change policy, immigration, economy, and environment. The White House disputed the need to combat climate change, legitimizing the US withdrawal from the Paris Agreement. Doubts were cast on all evidence presented by the scientific community for decades on global warming and its anthropogenic origin.

The Trump White House forcibly prevented migrants from crossing the border. It legitimized arresting and caging them. The Trump White House also legitimized the economic principle of mors tua vita mea (your death, my life), an isolationist and protectionist outlook that takes the form of import tariffs, trade wars, and an "us against them" mentality [29–31].

Textbox 6.1 History background in the US

(1898–1914)	The birth of a great power
(1914–1920)	The end of isolationism
(1921–1947)	The inter-war period and World War II
(1947–1991)	Cold War
(1991–2007)	US hegemony
(2008-present)	New powers

6.4 Economics

The United States, with 328,461,000 inhabitants, is one of the most populated countries in the world. There are 33 people per square kilometre. Between the years 2010 and 2019, natural increase has been the largest source of population growth in the nation. But in 2019, for the first time in at least four decades, it dipped below one million, falling to 957,000, and continuing the trend to fewer births and more deaths.

The international migration has also declined gradually each year since 2016, with an annual rise of 0.5% in 2019. In 2016 the population in the US increased by 1,047,000 and between 2018 and 2019 this number decreased to 595,000.

Therefore, the net international migration decrease during the last decade, together with the decline in natural increase of the population has caused a gradual slowdown in the United States' population growth [32, 33] (Fig. 6.1).

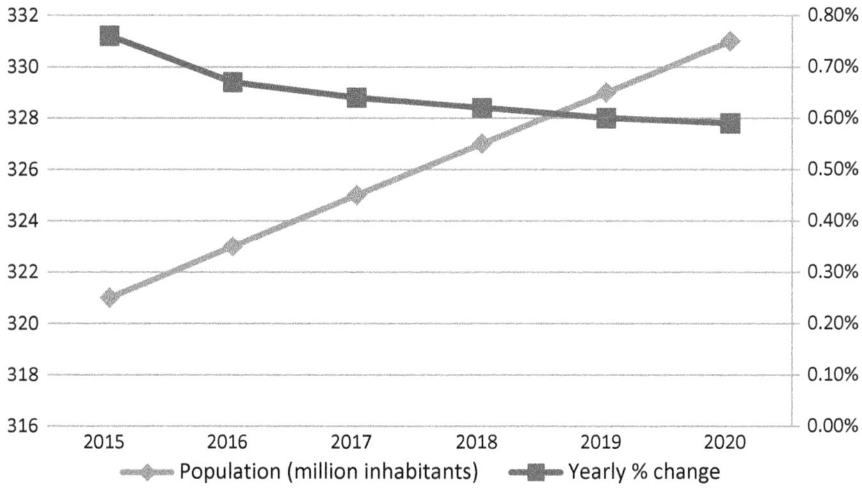

Fig. 6.1 Evolution of the population. *Source* own elaboration and adaptation from worldometers.info

The United States, in a very industrialised country, and the productivity level, is very high. The most relevant sectors are agriculture (soybeans, beef, cotton, and corn), machinery manufacturing, chemistry, food, and automation. The tertiary market is also very relevant mainly focused on finance, insurance, real estate, renting, and leasing. The agricultural sector is one of the greatest in the world, especially in California where approximately one-third of the country's vegetables and two-thirds of nuts and fruits are produced. However, the agriculture sector only accounts the 0.9% of GDP and employs 1% of the labour force [34, 35]. Considering also food and food related other industries, the primary sector shared 5% of the country's GDP, contributing USD 1.109 trillion in 2019 [36]. In the same year, around 22.2 million full or part-time jobs were related to the agriculture and food sectors [37, 38].

The industrial sectors, including a wide range of activities, share around 18% of the GDP and employs about 20% of the labour force. Among the industrial activities, the US is a world leader in the pharmaceutical and aerospace industries. The United States is also very rich in natural resources, and thanks to this abundance it also leads the production of a number of minerals. Apart from this, the country is also the world´s largest producer of natural gas liquids, aluminium, and nuclear energy, and the third largest oil producer. This variety of activities has allowed the US to maintain a diversified production.

Private goods-producing industries fell by 34.4% in the second quarter of 2020, following the COVID-19 pandemic. This huge drop largely reflects a decline in durable goods manufacturing (led by motor vehicles, bodies and caravans, and parts) [37, 39, 40].

The US economy depends mainly on the tertiary sector. The service-based sector shares approximately 77% of the GDP and employs more than 79% of the country's labour force. The tertiary sector in the US includes the insurance, finance, real state,

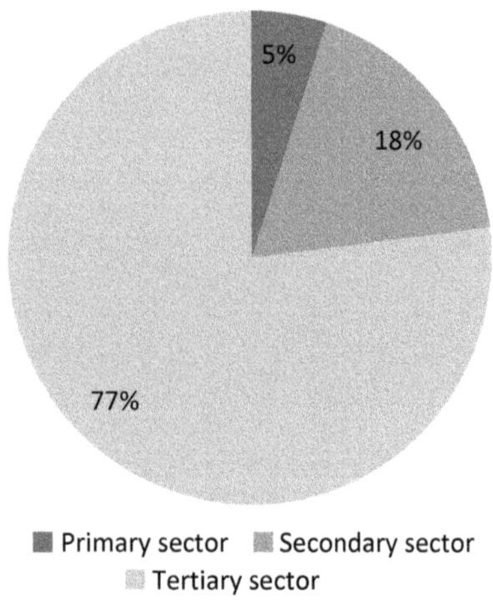

Fig. 6.2 US GDP by sector. *Source* Own elaboration and adaptation from https://www.worldbank.org/en/home

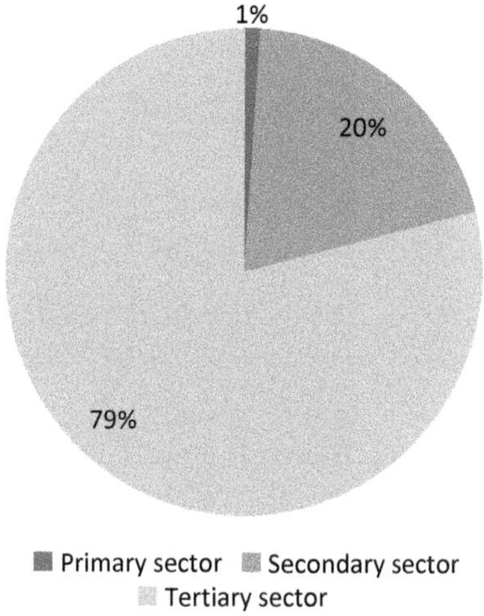

Fig. 6.3 US Labor Force by sector. *Source* Own elaboration and adaptation from https://www.worldbank.org/en/home

leasing, and renting sectors contributing 21% to the GDP, the professional and business service with 12.8% of the share, and government sector with a contribution of around 12.3% to the GDP in 2019 [41, 42] (Figs. 6.2 and 6.3).

Figure 6.4 shows the evolution of the per capita gross domestic product of the United States [43].

The United States is one of the highest energy consumers. The US citizens use energy for industry, businesses, homes, and transportation. Figure 6.5 shows the main sector where energy is consumed [44].

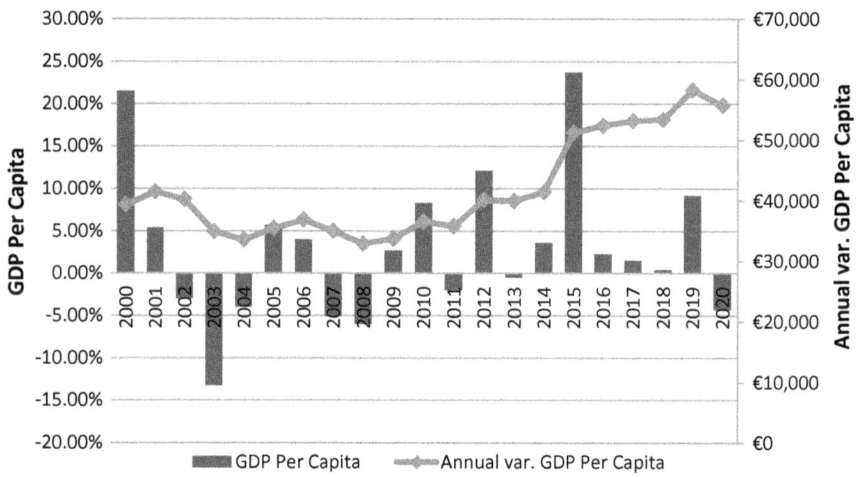

Fig. 6.4 Annual GDP per capita and annual var. GDP per capita. *Source* own elaboration and adaptation from Expansion/Datosmacro.com

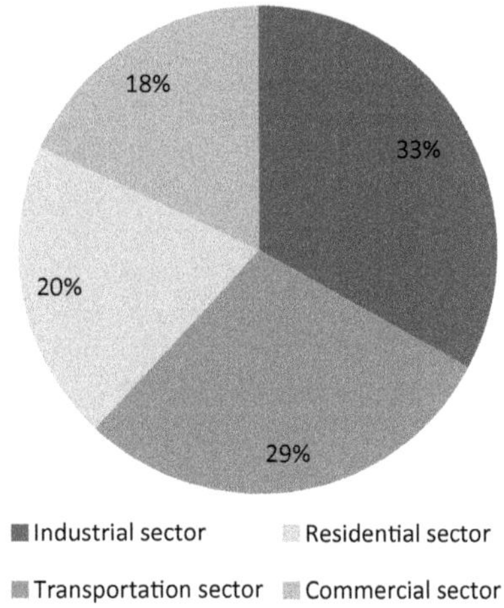

Fig. 6.5 Energy consumption by sector in the USA. *Source* own elaboration and adaptation from eia.gov

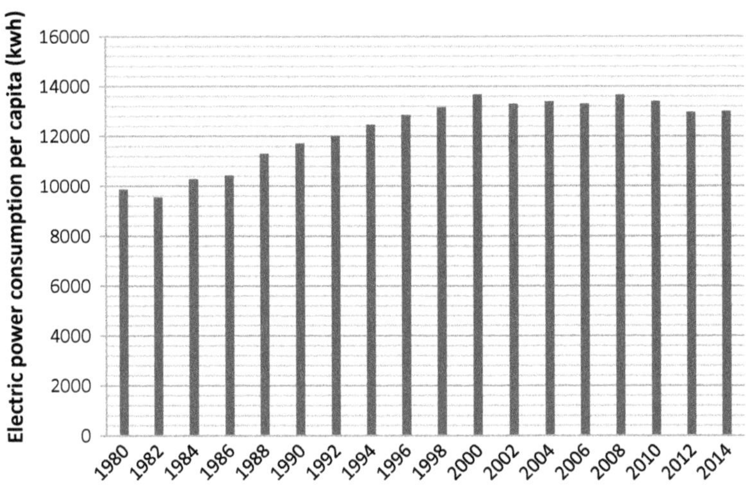

Fig. 6.6 USA's per capita electric power consumption. *Source* own elaboration and adaptation from Worldbank.org

Regarding the per capita electric power consumption, during the last years, there is a downward trend as can be seen in Fig. 6.6.

6.5 Energy Resources

The US is the world's largest producer of oil and the 14th largest by reserves; the third largest producer of coal and the largest by reserves (including by far the world's most profitable deposits in Wyoming and Montana), with about 27% of total production, about 491,000 tonnes; and, again, the largest producer of natural gas, where it is the third largest by reserves. Nature has blessed the US with unimaginable quantities of fossil fuels in Europe, and the country's entrepreneurial spirit and its ability to apply new technologies, such as the controversial 'fracking' in oil and gas extraction, have done the rest. The United States' large oil reserves are easily exploited thanks to relatively soft ground conditions. Early on, this resource allowed them to develop the civilian and military transportation industry, both on land, air, and sea [45, 46].

Over the last decade, the US has become one of the world's leading hydrocarbon producers, thanks to the development of shale drilling [47].

The proportion of total US energy production from fossil fuels had its maximum share in 1966 with the 93% of the production. Among the fossil fuels, petroleum products have been the main primary energy consumption source since 1950. Approximately 70% of these petroleum products were used for the transportation sector in 2018 [45, 47].

The second largest source of primary energy consumption in the US is natural gas. Its use has augmented considerably in the last decade, mainly due to the increase in natural gas-fired electricity generation.

In 1950, the largest amount of natural gas was used in the industrial sector (59%), while only 11% was aimed at the electric power sector. In 2018, the amount of natural gas used in the electric power sector is larger (35%) than in the industrial sector (34%).

Coal consumption had its peak in 2007, but its use has declined for the fifth consecutive year in 2018, with the lowest level since 1975. Even though in 1950, 47% of the coal was aimed at the industrial sector and 13% at the transportation sector, since 1960, almost all US coal has been used electric power sector. In 2018, the electricity generation sector accounted for 91% of the total US coal consumption [45, 48, 49].

Nuclear power is used in the electric power sector, with relatively stable quantities since mid-2000s. [50]

In 2019, the United States produced more energy than it consumed for the first time since 1957, and as a consequence, the US energy exports were higher than the imports for the first time since 1952 [45, 50].

The majority of US energy trade is from petroleum, which accounted for 69% of energy exports and 86% of energy imports in 2019. The majority of the imported crude oil is exported as petroleum after being processed in US refineries [45].

6.6 Energy Production

Both energy production and consumption reached record-high values in 2018. In 2019 energy production continued to grow by around 6% while consumption decreased by about 1%. As mentioned before, this was the first time the production exceeded the consumption on an annual basis since 1957. In 2020, the energy

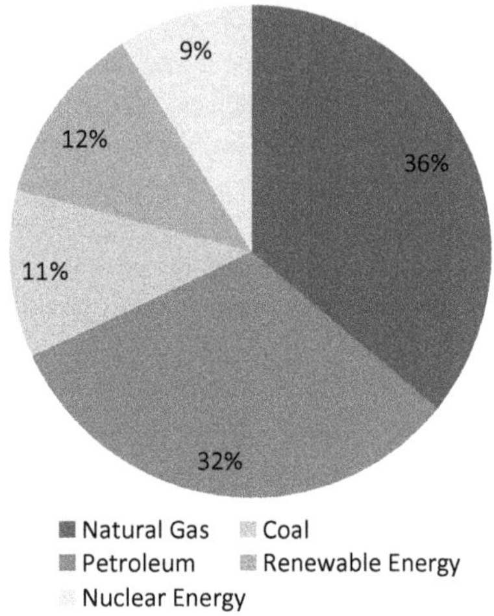

Fig. 6.7 Energy mix in the USA in 2020. *Source* own elaboration and adaptation from eia.gov

production decreased approximately 5%, but it was still 3% higher than consumption [45, 51].

Figure 6.7 shows the major primary energy production sources in the US in 2020. As can be seen, the dependency on fossil fuels (petroleum, natural gas, and coal), is notoriously high accounting for about 79% of the total production [45, 52].

Fossil fuels have been the main energy production and consumption for over 100 years, even though the energy mix has variated over time [45].

Coal consumption in the US reached its maximum value in 2007 and the production peaked in 2008. However, both production and consumption have decreased nearly every year since those peak years. The main reason is the decrease in the demand for coal for electric power generation. In 2020 both coal consumption and production had their lowest percentage share in the total energy consumption since 1949 and 1965 respectively [45, 53].

Natural gas production reached its highest amount in 2019, and it was around 2% lower in 2020. Regarding consumption, natural gas accounted the 34% of the energy mix in 2020. The annual natural gas production has been augmented thanks to more efficient drilling and production techniques which make it easier the extraction from shale and tight geologic formations. Therefore, natural gas production is higher than consumption since 2017. This increase in production has contributed to a reduction in natural gas prices, which in turn has increased the use of natural gas in the electricity generation and industrial sectors [45, 54, 55].

The US is also the world's largest oil producer and the ninth largest in terms of reserves. The importance of this hydrocarbon has skyrocketed since the application of large-scale fracking about a decade and a half ago, which has allowed it to almost double its reserves. Since 2017, the US has been exporting oil and oil derivatives [45].

Commercial nuclear energy generation in nuclear plants started in 1957, but it was not until 1990 that it started to grow. Its use levelled off after 2000. The number of operating nuclear reactors decreased in 2020 compared to 2000, but the amount of nuclear energy produced in 2020 was the second highest in record behind 2019, with 790 billion kWh. This is due to the combination of increased capacity of the power plants and shorter refuelling and maintenance cycles, which has helped to maintain a consistent level of annual nuclear electric power generation for the past 20 years [49, 55].

Renewable energy production and consumption reached their highest values in 2020, due to mainly record-high wind and solar energy production. In the case of hydroelectric energy generation, it was about 9% lower than the 50-year average. Biomass production and consumption was 10% lowest than in 2018 when it reached its highest value. Geothermal energy production and consumption in 2020 was almost the same as in 2014 when its use had the highest annual level recorded [56–58].

6.6.1 *Renewables*

In 2019, the US annual energy consumption from renewable energy sources was higher than the coal consumption for the first time since 1885 [45]. This reinforces the fact that coal consumption continues to decline, especially in electric power generation. Its consumption in 2019 was almost 15% lower than the consumption in 2018. The total renewable energy consumption increased by approximately 1% between those years.

Biomass was the most important energy source in the US until the mid-1800 and it was the only commercial-scale green energy source until the first hydroelectric power plant started to generate electric power around 1880. Renewable energy use started to increase in 2015, especially with the development of wind and solar energy

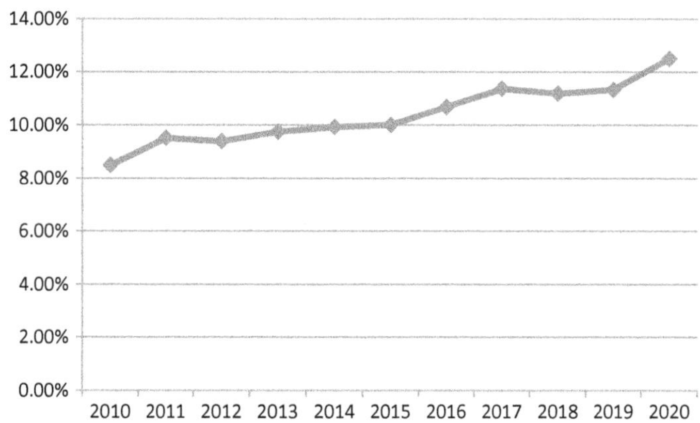

Fig. 6.8 Renewable energy share in the US primary energy consumption. *Source* own elaboration and adaptation from eia.gov

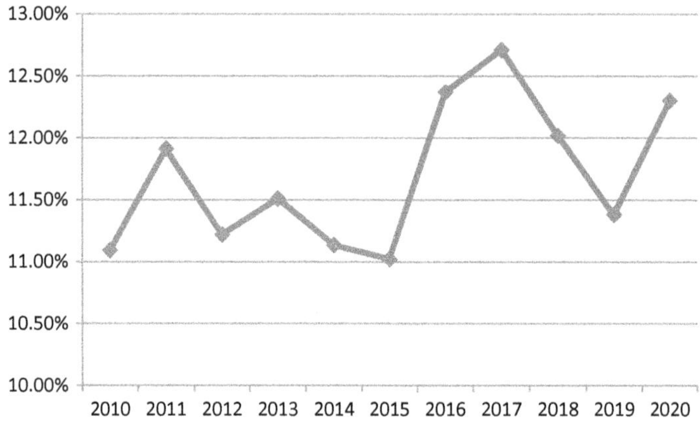

Fig. 6.9 Renewable energy share in the US primary energy production. *Source* own elaboration and adaptation from eia.gov

in the power sector. In 2019, electric power generation from wind energy overtook hydroelectric for the first time and currently, it is the most used renewable energy source for electricity generation in the US per year [56, 57].

In the context of serious economic crisis as a result of the coronavirus, renewables in the United States have grown to account for 20% of all electricity production in 2020.

The addition of new wind and solar capacity in unprecedented quantities in the United States, together with the potential of hydroelectric and nuclear power, accounted for 40% of total electricity production in the United States in 2020. All this in a year has also seen continued declines in coal-fired power generation [56, 58] (Figs. 6.8, 6.9 and 6.10).

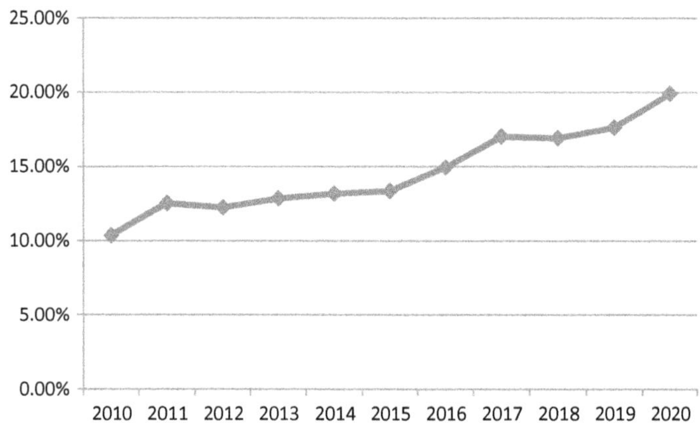

Fig. 6.10 Renewable energy share in the US electricity generation. *Source* own elaboration and adaptation from eia.gov

6.6.2 *Environmental Impact*

Globally, the United States is the largest energy consumer, with approximately 17% of the world's total, in line with its share of world GDP. Its emissions have been on a downward trend, albeit with ups and downs, since 2007. In absolute terms, it is the world's second largest emitter of greenhouse gases (GHG) after China. Adjusted for population, it is the second largest emitter after Luxembourg, and releases more than twice as much GHG into the atmosphere per capita as China.

For its part, total US emissions fell 20% below 2005 levels as a result of the pandemic. CO_2 emissions from road, rail, and aviation transport fell by 14% as a result of less travel, although it remains the largest emitting sector [59, 60] (Fig. 6.11; Table 6.1).

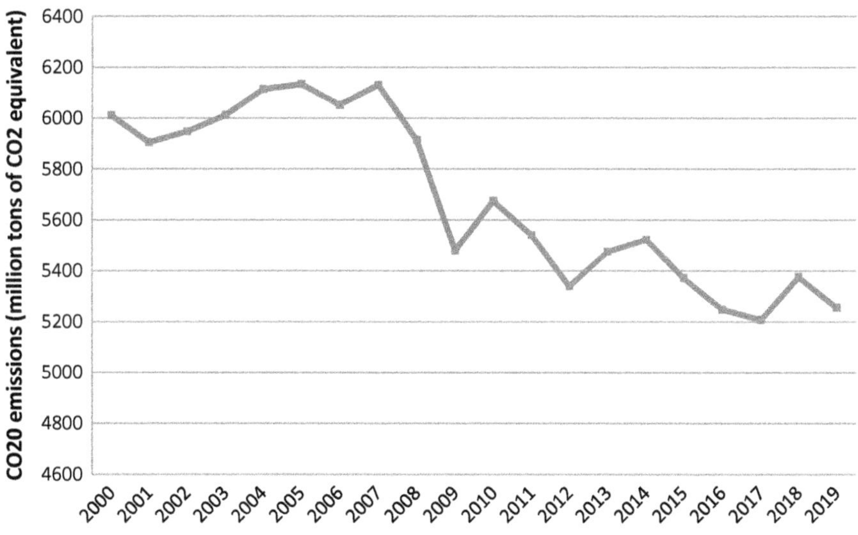

Fig. 6.11 CO_2 emissions (million tons of CO_2 equivalent). *Source* own elaboration and adaptation from cfpub.epa.gov

Table 6.1 GHG emissions by economic sector in 2019

Greenhouse gas emission by economic sector 2019 (%)	
Transportation	28.6%
Energy Generation	25.1%
Industry	22.9%
Agriculture	10.2%
Commercial	6.9%
Residential	5.8%

6.7 Energy Policy

The US's return to the Paris Agreement, under President Joe Biden, highlights the new administration's new approach to climate, with completely different policies to those pursued by the Donald Trump Administration.

The Biden administration's climate goals include halving the carbon footprint of the nation's building stock by 2035 and making drastic investments in energy efficiency in existing buildings, including four million retrofits [61].

In addition, he wants to have 100% emission-free electricity generation by 2035, as well as expand tax credits for clean energy and efficiency and install millions of solar panels and thousands of wind turbines on land and at sea [62, 63].

Biden has set a goal of eliminating pollution from fossil fuel use in the energy sector by 2035 and from the country's economy as a whole by 2050. To this end, the Democrat seeks to boost the growth of the clean energy sector, such as solar and wind power, and decrease dependence on oil and gas [64].

The main objective of his plan is to halt global warming caused by human action, which multiplies the impact of meteorological phenomena such as forest fires, which have intensified in recent years on the Californian coasts, as well as in the Amazon and many other places on the planet; and hurricanes, which are increasingly frequent and aggressive on the Atlantic coasts [62, 63].

Similarly, the new government wants to articulate a recovery from the coronavirus crisis based on the transition to clean energy and job creation, which is a very similar initiative to the one established by the European Union [65].

Biden's philosophy is the exact opposite of Trump's, and is based on three elements:

- Favouring the replacement of fossil fuels (coal, natural gas, and oil) with renewables, especially solar and wind, to achieve a zero emissions target in the energy sector by 2035 and across the economy by 2050 (halfway between 2035, which is the EU's, and 2060, which is China's)
- The transition of employment from these sectors to others, and the creation of jobs in renewable energy, with an ambitious but ambiguous plan to invest two trillion dollars (1.85 trillion euros) in this goal
- Coordinating US energy policy with other countries to reduce global warming gas emissions.

For the Democratic left wing, which is particularly strong among the younger, more urban electorate, climate change and energy policy is an overriding issue. For the Democratic left, the fight against climate change is a tool to combat income inequality and create economic opportunities for ethnic minorities. There is the concept of the 'Green New Deal'8, a concept that takes its name from the 'New Deal' with which President Franklin Delano Roosevelt developed public spending and the state to combat the Great Depression of the 1930s, and also the idea of 'environmental justice'. But the coalition that brought Biden to the White House is complex, and also includes middle-class voters in Pennsylvania—where coal and, more recently, fracked natural gas plays an economically important role—and unions, whose priority is defending the interests of their members. Consequently, Biden cannot take an overly belligerent stance on energy, as he risks damaging his and his party's electoral base unless he is able to ensure that the jobs lost in fossil fuels are replaced by jobs in renewables or other sectors [66, 67].

A second constraint is that the Democratic majority in Congress is very slim. This makes it difficult to pass legislation that changes the rules of the energy game, although the president's ability to act simply by using regulatory agencies is very large. The problem with that policy is that it can be reversed by his successor, just as Biden has already reversed many of Donald Trump's decisions. It is ruled out that Biden will alter the fossil fuel tax regime (e.g. by raising gasoline taxes), and it seems clear that he will present his measures as a way to strengthen US competitiveness and energy independence (the latter, which the US already is).

The importance of the energy debate is illustrated by the steps Biden took immediately after taking office [68, 69]:

- The return of the United States to the Paris Agreement;
- Temporarily banning a number of oil exploration projects in the Alaskan Wildlife Refuge;
- The suspension of the issuance of new licenses to exploit oil and gas through fracking on federally owned land, which also includes the 200-mile Exclusive Economic Zone off the US coast. Permitting will not be reactivated until a review of the projects' impact on global warming and ocean acidification is completed. There is no date for the conclusion of that review [70];
- The cancellation of the authorisation for the construction of the Keystone XL, a variant of the Keystone pipeline, a network of pipelines that carries oil from the tar sands of Fort McMurray in Alberta, Canada, to refineries in the US Gulf of Mexico [71, 72].

These measures are, in the short term, symbolic in scope. Re-entry into the Paris Treaty does not entail any limit on emissions per se, since that is left to be negotiated by each country. The same applies to the decision to suspend the issuing of fracking permits because, while it is true that unconventional wells accounted for 90% of those drilled in 2018, according to the Baker Hughes consultancy, only 11% of fracking permits are on federal land. Moreover, with the collapse in crude oil prices over the past year, demand for new permits has plummeted. In the long term, however, these decisions can have a very considerable impact. Indeed, the scope of Biden's Executive Orders on energy has come as a negative surprise to the hydrocarbons sector.

The president has other mechanisms at his disposal to carry out his policies, notably through regulatory mechanisms and also through foreign action, as long as it does not require the approval of international treaties, for which Senate approval is necessary. Some of these elements will have a clear impact on business activities, for example in sectors such as construction, infrastructure, transport, and agriculture.

These measures include [73, 74]:

- The setting of an emissions target for 2025, as a first step towards the emissions neutrality mentioned above;
- The consideration of the origin of energy among the criteria for public procurement;
- The development of the high-voltage power grid to bring renewable energy from central states such as Texas and Nebraska (both Republican) to the coasts (Democratic), where demand is greatest;
- Working with other countries—including China, the US's main geostrategic rival and also the world's biggest emitter—to find ways to limit emissions. In the specific case of China, India, and other developing countries, Washington's focus will be on reducing fuel subsidies in those economies;
- Tightening controls on the emission of methane, a gas that has a far greater effect on global warming than carbon dioxide and often leaks in quantities that are difficult to estimate in fracking operations;
- The provision of more public resources for the development of new nuclear reactor technologies;
- Strengthening energy efficiency criteria in buildings and in particular in public buildings.

The president will use the three billion dollars (almost 2.5 billion euro) post-Covid-19 economic recovery plan he plans to unveil in March to introduce subsidies and public support for renewables, as well as rehabilitation for other uses of land used by coal mining companies. With this policy, Biden will try to push through his reform by codifying it into law.

6.8 Wind Turbine Installations

Wind power became the top US choice for adding new power after building 9.1 gigawatts (GW), accounting for 39% of new utility-scale power additions in 2020. With these additions, currently, there is over 105 GW of wind power installed capacity in the United States; this makes the US the second largest country in the world in installed capacity [75]. There are over 56,800 wind turbines operating in 41 states, with a power capacity of 105,583 MW [76]. This would be enough to power more than 32 million homes in the United States. Wind energy is currently the largest supplier of renewable energy in the US, it accounted for 7% of the country's electric power in 2019. In six states (Iowa, Kansas, Maine, North Dakota, Oklahoma, and South Dakota) wind energy provides more than 20% of the electric power generation, and in Iowa and Kansas, currently, wind energy is the main source of electricity generation, with the 40% of their electric power generated from wind energy in 2019. In 14 US states, 10% of the electricity is generated with wind power, and in another 20 states, 5% of the electric power comes from renewable energy sources [75–78] (Fig. 6.12).

Wind energy project makers developed 55 projects in 19 US states in 2019. Iowa and Texas installed individual record amounts of wind energy in a single year,

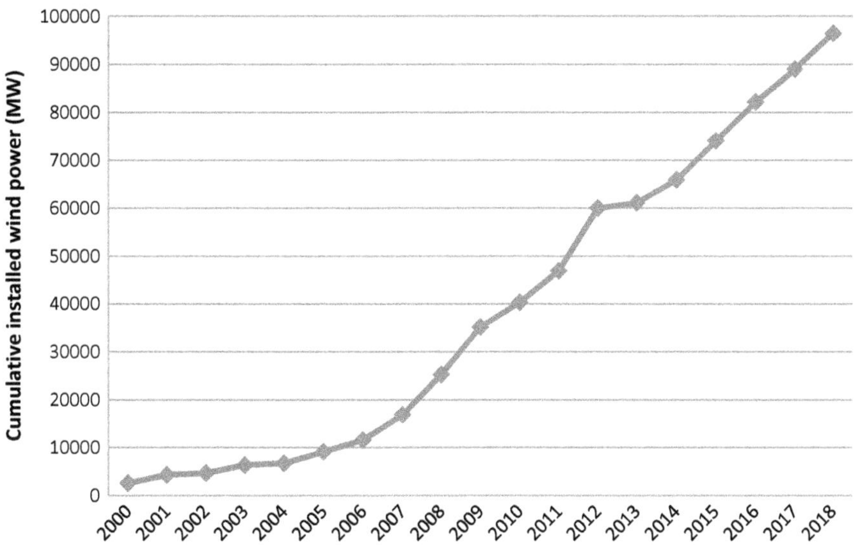

Fig. 6.12 Cumulative installed wind power capacity 2000–2018. *Source* own elaboration and adaptation from https://es.statista.com

leading the country in wind power additions. South Dakota, after installing 506 MW and increasing its installed wind capacity by over 50%, had the highest percentage increase in 2019 [75, 79]. Having ended the last decade with record growth, the wind industry is poised for continued strong expansion in the coming years. A further 44 GW of capacity is under construction or in advanced stages of development.

Along with growth in the onshore wind sector, many states are looking to take advantage of the opportunity for rapid expansion of offshore wind. East Coast states aim to develop more than 25 GW of offshore installed capacity before 2035. To help in the achievement of this objective, states and utilities selected 4404 MW of offshore wind capacity through state-issued applications in 2019 [75].

Wind energy is not only the largest source of renewable electric power in the US; it also provides several environmental benefits. In 2019, thanks to wind energy, carbon dioxide emissions decreased by 198 million metric tonnes, the equivalent of removing 42 million cars from the road. Wind projects also saved 390 billion

Table 6.2 Top 5 wind energy states in the United States

State	Installed capacity (MW)
Texas	28,843
Iowa	10,190
Oklahoma	8172
Kansas	6128
California	5973

litres of water last year and helped to improve the air quality by reducing sulphur dioxide emissions by 232,000 metric tons and nitrogen oxides by 168,000 metric tons [76, 78].

The top 5 wind energy states in the United States are the following [78, 80] (Table 6.2).

Texas

Texas leads the country's wind power generation with 28,843 MW of total installed capacity in 2019. This capacity was higher than coal as an energy source for the first time in 2019. During the tenure of Governor Rick Perry, wind energy increased from 116 MW to more than 11,000 MW, and investments were made to connect the windy and empty West Texas to cities where additional power was needed.

The state's largest wind energy facility is the 781.5 MW Roscoe wind farm, which was developed by Mitsubishi, Siemens AG, and General Electric.

The 781.5 MW Roscoe wind farm is the state's largest facility with 627 turbines, delivered by Mitsubishi, Siemens AG, and General Electric. This facility is owned and operated by E.ON Climate and Renewables and it occupies 400 km2 of land.

Iowa

The second largest nation in wind energy installations, with a total installed capacity of 10,190 MW in 2019, is Iowa. This state has the highest percentage of wind-produced power in the US, with a total of 41.7% in 2019. Important wind turbine manufacturers such as Siemems Gamesa or TPI Composites, have made significant investments in Iowa.

According to the American Wind Energy Association, there were as many as 116 wind projects online in the last quarter of 2019, with 1175 MW of wind capacity under construction and 384 MW in advanced development.

Oklahoma

Oklahoma ranks third in the wind installed capacity list, with 8172 MW of installed capacity. It has the potential to provide almost 10% of the electric power requirements in the United States. In 2016, wind power generated more electric power than coal for the first time, and in 2019 renewable energy sources generated 31.7% of all the electricity produced in the state. The wind industry in Oklahoma generates approximately 9,000 direct and indirect jobs.

The largest wind farm in the state is Blue Canyon, with an installed capacity of 423.45 MW, enough to power approximately 104,000 homes per year.

Kansas

In May 2009, the Kansas Legislature ratified the Renewable Energy Standards Act, which required utilities to generate or purchase 20% of their energy from renewable sources by 2020. The current installed wind capacity in Kansas is three times higher than the one in early 2008. It has also been established a 10-year property tax exemption for sites used for electricity generation from renewable sources. It has been estimated that the Kansas power system can offer around 7,000 MW for wind energy exports by 2030.

California

Large wind farms were first developed in California, beginning in the early 1980s. Currently, wind energy is the largest source of renewable energy in the state. The largest wind farm in the US, and the third largest of the world, is also located in California, the Alta Wind Energy Center with an installed capacity of 1550 MW. In September 2018, California Governor Jerry Brown signed two laws with the objective of providing 100% renewable electricity and achieving carbon neutrality by 2045.

6.8.1 Offshore Wind Energy

Currently, the offshore wind energy sector in the US is small compared to European developments. Just at the end of 2016, the first of its kind, the Block Island wind farm off the coast of Rhode Island, a facility operated by Ørsted and with a capacity of 30 megawatts and five turbines, went into operation.

According to Thomas Brostrøm, CEO of Ørsted North America Offshore, there are a number of states that are looking to increase their offshore capacity, including New York, which is targeting nine gigawatts by 2035, New Jersey, where officials are aiming for 7.5 GW by 2035, and Virginia, which is targeting 5.2 GW by 2034 [81].

The potential for offshore wind in the US is significant, with the Department of Energy (DOE) stating that offshore wind in the US has "a technical resource potential" of more than 2000 GW [82].

For Wood Mackenzie, there is strong support from many states along the east coast, and investment in offshore wind is expected to increase strongly. Projects with around nine gigawatts of capacity are already contracted or will soon be approved.

The greatest potential and opportunities are primarily in the northeastern United States, where state policies and favourable conditions are driving the industry, however, over time the southeast and west coast are also expected to become home to the construction of large offshore wind projects.

There is undoubtedly room for a significant increase in offshore wind in the US, it clearly has a long way to go to reach more mature markets such as Europe.

With effective policy intervention, such as an extension of the Investment Tax Credit, new state renewable energy mandates, and support for an offshore transmission backbone, around 34 GW could be added [83].

6.9 Solar PV Installations

According to the Solar Energy Industries Association (SEIA), the US solar industry grew 43% and installed a record 19.2 gigawatts (GWdc) of capacity in 2020. Specifically, total installed capacity in the first quarter of 2020 amounted to 3.6 gigawatts (GW), almost a gigawatt above any previous quarterly record, with residential, commercial, and industrial projects contributing almost 1.6 GW to that total. More than 56% of the total capacity added during this quarter was connected to the grid in January and Florida was the most active state in this period, with 596 megawatts (MW) installed compared to 359 MW in California [45, 84, 85].

It is expected that the US solar industry will install a cumulative 324 GWdc of new capacity to reach a total of 419 GWdc over the next decade [45, 86].

There are two reasons why such growth will be experienced:

- Robust demand from utilities and corporations looking to meet greenhouse gas reduction targets
- The falling costs of the technology that have underpinned the market for solar home installations

The US as a whole had cumulative utility-scale solar power capacity of close to 41 GW on 31 March 2020, 18.4% more than on the same day in 2019 [87].

However, SEIA's latest reviews of the country's sector performance have slightly reduced the growth expectations set last year due to the impact the coronavirus crisis is having on all economic activities. Thus, they have lowered the forecasts for the period between 2020 and 2025 to 113 GW [45, 87].

41 states apply different variations of net balancing policies to regulate solar self-consumption, although the evolution of distributed generation regulatory policies is following a different path in each of them.

In Nevada, as a result of pressure from utilities, the regulator has changed the remuneration for self-consumption from net balancing to market price, even though a study concluded that net balancing for systems installed between 2004 and 2016 would provide a benefit to non-self-consuming domestic consumers of USD 36 million ($36 million) over the lifetime of the systems. This decision follows the federal government's extension of investment tax credits for solar projects until 2020.

Arizona and Hawaii have also eliminated net balancing, but have introduced other tariffs and regulations to remunerate self-consumption. Hawaii has established two tariffs: grid-supply and self-supply. The former offers a slightly lower remuneration than the domestic tariff, while the latter does not allow injection into the grid, but also does not introduce charges for self-consumed energy. Arizona, on the other hand, will remunerate surpluses on the basis of historical PPA prices and cost analysis, with no charges for self-consumed energy and no retroactive measures.

Other states, however, continue to support net balancing: California has extended it until at least 2019 and New York is in the process of maintaining it for domestic self-consumers and small businesses until 2020. Massachusetts has extended its Solar Renewable Energy Credit programme for this year.

Fortunately, cost–benefit analyses have become a fundamental part of state tariff-setting processes for self-consumption. Most of these analyses, whether conducted by regulators, national laboratories, or universities, conclude that the economic benefits of net balancing outweigh the costs and that, in general, it is a net benefit for distribution companies and domestic consumers.

Studies by state regulators conclude that solar energy provides a substantial public benefit because:

- Reduces electricity prices, due to the displacement of more expensive energy sources
- Reduces air and climate pollution
- Reduces grid system costs
- Reduces the need to build more power plants to meet peak demand
- Stabilises prices and promotes energy security.

These avoided costs represent a net benefit to household consumers who are not self-consumers.

In terms of methodology development, in 2014, the Minnesota Public Utilities Commission approved the first federal "value of solar" methodology that asserted that distributed solar generation is worth more than the price paid for it. Also, the National Association of Regulatory Utility Commissioners (Naruc) has published the Distributed Resource Compensation Handbook, a guide to different models for remunerating distributed generation with a focus on the compensation that distribution companies should receive. In no case does the model consist, as in Spain, of paying charges for self-consumed energy.

Other analyses with a more long-term focus come to the same conclusion: the target metric is generally a net benefit for the grid and all taxpayers [45].

In 2021 the United States plans to complete more than three dozen solar projects with capacities exceeding 100 MW. The 420 MW Permian Energy Center is the largest of these, which has been developed in Andrews County, Texas, by Denmark's largest energy supplier and wind project developer, Ørsted [88]. The project is northwest of Midland and is located in the heart of the Permian Basin oil and natural gas field of West Texas. The Permian solar plant includes more than 1.3 million photovoltaic panels as well as a 40 MW one-hour battery energy storage system [89].

Two 300 MW solar projects in Ohio and Wisconsin are the next largest installations expected to be commissioned in 2021. Hecate Energy Highland is being developed in Highland County, Ohio, east of Cincinnati. Hecate Energy is a Chicago-based solar and energy storage developer [88].

The Badger Hollow solar farm is being developed by Chicago-based Invenergy and is located in Iowa County, Wisconsin, west of the state capital, Madison. Work is underway and that facility is also scheduled to enter service in December 2021 [89].

In total, more than 9.7 GW of solar capacity is expected to be completed in 2021. Approximately one-third of that capacity is in Texas. The state's 15 projects have an average size of 195 MW [90].

Apart from these new projects, four of the 20 larger solar PV plants in the world are located in the United States.

Solar Star Solar Farm I y II. 597 MW

Solar Star is a 597 MW photovoltaic power plant located near Rosamond, California. It consists of two phases: the first phase of 318 MW and the second phase of 279 MW. The plant was finished in 2015 and it is currently, in terms of installed capacity, the fourth-largest plant in the world. It occupies an area of around 13 km^2 and has around 1.7 million solar panels of high efficiency panels which are mounted on single-axis trackers.

It is worth mentioning that this plant uses a smaller number of panels (with higher efficiency) and therefore a smaller area compared to other similar installations such as Desert Sunlight or Topaz Solar Plants, which use a larger number of cadmium telluride PV modules instead of conventional crystalline silicon PV panels [91, 92].

Copper Mountain. 552 MW

The Copper Mountain Solar PV plant, located in Nevada, is the second largest plant in the US with 552 MWp of total installed capacity. It has been built in four phases and its annual production I 1000 GWh. The energy generated from phase 1 was sold to Pacific Gas & Electric under a 20-year power purchase agreement (PPA). The power produced from phase 2 was also sold to the same company with a 25-year power purchase agreement (PPA). Power generated from phases 3 and 4 is sold to the Southern California Public Power Authority.

Phase 3 of the project is one of the largest photovoltaic developments ever undertaken in the USA, with an installed capacity of 250 MW, which, together with the two previous phases and the 94 MW of the fourth phase, gives a total of 552 MW [92, 93].

Desert Sunlight Solar Farm. 550 MW

The Desert Sunlight PV plant is located in the Mojave Desert, California, and has an installed capacity of 550 megawatts (MWAC). It occupies more than 16 km^2 of land in the vicinity of Desert Centre.

The plant has approximately 8.8 million cadmium telluride modules from US-based First Solar's thin film manufacturing facilities. The plant was inaugurated in February 2015, with the same installed capacity (550 MW) as Topaz Solar Farm, making the two plants tied for sixth place in the ranking of the largest operating solar plants by installed capacity in the world.

Construction of the project was carried out in two phases, both of which were supported by long-term power purchase agreements (ppa). Phase 1 has a capacity of 300 MW, the output of which is sold to the Pacific Gas & Electric Company. Phase 2 with a capacity of 250 MW, was sold to Southern California Edison. The development of the project generated more than 550 jobs in Riverside County, California during construction [91].

Topaz Solar Farm. 550 MW

MidAmerican Solar commissioned the world's largest and most powerful solar power plant to date, Topaz Solar Farm, in San Luis Obispo, California, in 2014. The plant covers an area of 26 km^2 and houses a total of 9 million First Solar photovoltaic panels with an output of 550 MW.

The $2.5 billion plants is capable of powering a total of 160,000 homes, saves a total of 377,000 tonnes of CO_2 emissions per year, equivalent to the equivalent of 73,000 vehicles on the road. Operations and maintenance of Topaz will continue to be handled by First Solar, and production will go entirely to Pacific Gas & Electric, with which the company signed a 25-year ppa contract. What was the world's largest PV plant two years ago now shares sixth place with Desert Sunlight [92, 94].

6.10 Solar Thermal Installations

The US solar market is growing rapidly as the industry expands innovative new technologies and reduces manufacturing and installation costs. Public interest in clean technologies and renewable energy has increased demand for solar power. The movement towards a sustainable energy infrastructure has encouraged consumers and businesses to demand renewable energy sources, such as solar power.

The United States and Spain also share 90% of the Top 10 places in the world's largest solar thermal plants. The United States leads the ranking, with five plants in the Top 10, occupying the top five places.

Unlike photovoltaics, which converts solar radiation directly into electricity, solar thermal installation generates heat. More than 300,000 mirrors arranged in concentric circles around the three towers reflect and concentrate the solar radiation received onto a boiler at the top of each tower, raising the temperature of the water inside to more than 530 °C. This heat turns the turbines, which generate electricity.

The Mojave Desert has ideal climatic characteristics for this type of facility, as it receives sunlight 330–350 days a year, and its elevation and dry air enhance these favourable conditions. Praised for its engineering, the plant has also been widely criticised for its environmental impact [45].

Ivanpah. 392 MW

Three huge solar farms in the middle of the Mojave Desert, shared by the states of Nevada and California, make up the Ivanpah complex, the world's largest solar thermal plant, which occupies an area of 13 km^2 just 60 km south of Las Vegas.

At full capacity, its three 139-m-high towers and more than 300,000 computer-controlled mirrors can produce 392 MW, a clean supply equivalent to reducing 400,000 tonnes of carbon dioxide per year or, in other words, the equivalent of taking 72,000 vehicles off the road. A contribution that will help the state of California get closer to its goal of obtaining 33% of its electricity from renewable sources by 2020.

Ivanpah is owned by NRG Energy, BrightSource Energy, and Google, and is so large that it alone provides 30% of all solar power generated in the United States [95, 96].

Solar Energy Generation Systems (SEGS). 354 MW

Solar Energy Generating Systems (SEGS), with an installed capacity of 354 MW is located in the Mojave Desert in California. It generates 662 GWh of energy per year.

This project combines nine solar installations, divided into three sections: SEGS I-II (44 MW), SEGS III-VI (150 MW), and SEGS VIII-IX (160 MW). SEGS I-II first began operation in 1986, SEGS III-VII in 1988, and SEGS VIII-IX in 1990. Sunray Energy owns the SEGS I-II facilities, while NextEra Energy Resources owns and operates the remaining facilities, with all power generated being distributed by Southern California Edison Electric Company [97].

Solana. 280 MW

The Solana Generating Station is a solar thermal plant near Gila Bend, Arizona, about 110 km southwest of Phoenix, completed in 2013. Built by Spanish company Abengoa Solar, it has a total capacity of 280 megawatts (MW) gross, from two 140 MW (net 125 MW) gross steam turbine generators, which is enough to power 70,000 homes while avoiding around 475,000 tonnes of CO_2 per year.

The plant employs a unique Concentrating Solar Power (CSP) technology developed by Abengoa and covers an area of 7.8 km^2. Arizona Public Service (APS) entered into a long-term contract to purchase 100% of the power generated at the Solana plant to meet the Arizona Corporation Commission's (ACC) mandate that regulated utilities provide 15% of their electricity from renewable energy by 2025 [98, 99].

Mojave Solar. 280 MW

The 280 megawatts (MW) Mojave Solar plant, built by Abengoa, is located in the Mojave Desert, 90 km northeast of Los Angeles, near Barstow (California, USA). Mojave Solar produces clean energy capable of supplying approximately 91,000 homes in the state of California and will prevent the emission of about 223,440 tonnes of carbon dioxide (CO_2) annually into the atmosphere, according to the company.

The electricity generated at Mojave will be supplied to Pacific Gas & Electric Company for the next 25 years under an existing power purchase agreement (PPA). The project also contributes to California's goal of achieving 33% of total electricity production from renewable sources by 2020.

During its construction phase, the solar plant has created more than 2,200 jobs and currently employs around 70 people permanently in operational and maintenance activities [100, 101].

Genesis. 250 MW

The Genesis Solar Energy Project is a 250-megawatt parabolic trough solar thermal power (CSP) plant sponsored by NextEra Energy Resources, located in the Mojave Desert in California.

The solar power plant consists of two separate 125 MW sections and covers an area of 5.5 km^2. The plant has an annual production capacity of 560 GWh, which is supplied to Pacific Gas & Electric through a long-term power purchase agreement [102].

6.11 Conclusions

Globally, the United States is the largest energy consumer, with approximately 17% of the world's total, in line with its share of world GDP. Its emissions have been on a downward trend, although with ups and downs, since 2007.

The US's return to the Paris Agreement, under President Joe Biden, highlights the new administration's new approach to climate, with completely different policies to those pursued by the Donald Trump Administration.

Wind energy is currently the largest supplier of renewable energy in the US, it accounted for 7% of the country's electric power in 2019. Along with growth in the onshore wind sector, many states are looking to take advantage of the opportunity for rapid expansion of offshore wind. There is undoubtedly room for a significant increase in offshore wind in the US, it clearly has a long way to go to reach more mature markets such as Europe.

The United States and Spain also share 90% of the Top 10 places in the world's largest solar thermal plants. The United States leads the ranking, with five plants in the Top 10, occupying the top five places.

The conclusions are quite clear and can be applied to any country that wants to develop a sustainable solar industry to benefit society to the full extent of its advantages: there is a need to discuss the most efficient way to integrate distributed generation into the grid, based on rigorous analysis that captures all economic, social and environmental benefits, with a view to establishing a fair cost recovery strategy that does not pose a threat to solar energy.

References

1. State Area Measurements and Internal Point Coordinates, January 1, 2020. https://www.census.gov/geographies/reference-files/2010/geo/state-area.html
2. https://peakvisor.com/adm/usa.html
3. Hupp CR (2000) Hydrology, geomorphology and vegetation of Coastal Plain rivers in the south-eastern USA. Hydrol Process 14:2991–3010
4. Dougherty E, Morrison R, Rasmussen K (2021) High-resolution flood precipitation and streamflow relationships in two US river basins. Meteorol Appl 28:e1979
5. Woodhouse CA, Wise EK (2020) The changing relationship between the upper and lower Missouri River basins during drought. Int J Climatol 2020(40):5011–5028
6. National Centers for Environmental Information. August 2021. https://www.ncdc.noaa.gov/cag/national/time-series/110/

7. Shaw RH (1982) Climate of the United States, in handbook of soils and climate in agriculture. CRC Press, ISBN, p 9781351073073
8. Bhandari S, Kalra A, Tamaddun K, Ahmad S (2018) Relationship between ocean-atmospheric climate variables and regional streamflow of the conterminous United States. Hydrology 5(2):30. https://doi.org/10.3390/hydrology5020030
9. Latner TA (2018) Cuban revolution in America: Havana and the making of a United States left. The University of North Carolina Press, ISBN 1469635461
10. Loyd JM, Mountz A (2018) Boats, borders, and bases: race, the cold war, and the rise of migration of the United States. University of California Press, ISBN, p 0520287975
11. Terrence Hopmann P, King TD (1980) From cold war to Détente: the role of the Cuban Missile Crisis and the partial nuclear test ban treaty, in change in the international system. Routledge ISBN 9780429052187
12. Immerman RH, Engel JA (2020) Fourteen points for the twenty-first century: a renewed appeal for cooperative internationalism. The University Press of Kentucky, ISBN 978-0-8131-7902-5
13. Desenrode S (2018) World War :1 the great war and its impacts. Aalborg University Press, ISBN 978-87-7210-004-3
14. Garon S (2020) On the Transnational destruction of cities: what Japan and the United States learned from the bombing of Britain and Germany in the second world war. Past Present 247(1):235–271. https://doi.org/10.1093/pastj/gtz054
15. Jackson S, O'Malley A (2018) The institution of international order: from the league of nations to the United Nations. Routledge, ISBN 978-1-138-09150-4
16. Woloch I (2019) Postwar moment. Yale University Press, ISBN 978-0-300-12435-4
17. Patrick Cottrell M (2018) The league of nations: enduring legacies of the first experiment at world organization. Routledge, ISBN 978-1-138-93088-9
18. McKinney KE, Sagan SD, Weiner AS (2020) Why the atomic bombing of Hiroshima would be illegal today. Bull Atomic Sci 76(4):157–165. https://doi.org/10.1080/00963402.2020.1778344
19. Fathyasyifa Shabrina I, Ketut Surajaya I (2019) Why Hiroshima as the target of US atomic bomb in 1945. Int J Innov Sci Res Technol ISSN No: 2456-2165
20. Kort M (2018) The Vietnam war re-examined. Cambridge University Press, ISBN 978-1-107-04640-5
21. Levy PB (1999) Blacks and the Vietnam war, in the impact of race on U.S. foreign policy. Routledge, ISBN 978-1-003-05915-8
22. Itzkowitz Shifrinson JR (2020) Eastbound and down: the United States, NATO enlargement, and suppressing the Soviet and Western European alternatives, 1990–1992. J Strategic Stud 43:6–7, 816–846. https://doi.org/10.1080/01402390.2020.1737931
23. Leffler M (2017) Strategy, diplomacy, and the cold war: the United States, Turkey, and NATO, 1945–1952. Safeguarding Democratic Capitalism: U.S. Foreign Policy and National Security, 1920–2015. Princeton University Press, Princeton, pp 164–186. https://doi.org/10.1515/9781400888061-007
24. Thomas C (2019) Afghanistan: background and U.S. Policy. Congressional research service
25. Dobbins J, Campbell JH, Mann S, Miller LE (2019) Consequences of a Precipitous U.S. Withdrawal from Afghanistan. RAND Corporation (2019) Stable. http://www.jstor.com/stable/resrep19894
26. Butt AI (2019) Why did the United States Invade Iraq in 2003? Secur Stud 28(2):250–285. https://doi.org/10.1080/09636412.2019.1551567
27. Yetiv SA (2018) The Iraq war of 2003. Why did the United States decide to invade? Routledge, ISBN 978-0-429-48904–4
28. Dodge T (2007) The causes of US failure in Iraq. Routledge, ISBN 978-0-429-33314-9
29. Woolhandler S et al (2021) Public policy and health in the Trump era. The Lancet 397(10275):705–753
30. McIntosh J, Mendoza-Denton N (2020) Language in the Trump era: scandals and emergencies. Cambridge University Press, ISBN 978-1-108-84114-6

31. Bomberg E (2017) Environmental politics in the Trump era: an early assessment. Environ Pol 26(5):956–963. https://doi.org/10.1080/09644016.2017.1332543
32. Vespa J, Armstrong DM, Medina L (2018) Demographic turning points for the United States: population projections for 2020 to 2060. Current Population Reports P25-1144
33. Roberts AW, Ogunwole SU, Blakeslee L, Rabe MA (2018) The population 65 years and older in the United States: 2016. American Community Survey Reports
34. The World Bank (2021) https://www.worldbank.org/en/home
35. Olmstead AL, Rhode PW (2017) A history of California agriculture. Giannini Foundation of Agricultural Economics
36. U.S. Department of Agriculture. https://www.usda.gov/
37. Nordea. https://www.nordeatrade.com/en
38. Tran TK, Sato H, Namatame A (2017) Key economic sectors and their transitions: analysis of world input-output network. In: Kreinovich V, Sriboonchitta S, Huynh VN (eds) Robustness in econometrics. Studies in computational intelligence, vol 692. Springer, Cham. https://doi.org/10.1007/978-3-319-50742-2_23
39. Bureau of Economic Analysis, U.S. Department of Commerce. https://www.bea.gov/
40. Luh S, Budinis S, Giarola S, Schmidt TJ, Hawkes A (2020) Long-term development of the industrial sector—case study about electrification, fuel switching, and CCS in the USA. Comput Chem Eng 133:106602, ISSN 0098-1354. https://doi.org/10.1016/j.compchemeng.2019.106602
41. U.S. Department of Commerce. https://www.commerce.gov/
42. Yum S (2020) The relationship between creative industries and the urban economy in the USA. Creative Ind J 13(2):95–116. https://doi.org/10.1080/17510694.2019.1668741
43. Expansión, Datos Macro. https://datosmacro.expansion.com/
44. American Geoscience Institute. https://www.americangeosciences.org/
45. U.S. Energy Information Administration. https://www.eia.gov/
46. Smith JL, Lee TK (2017) The price elasticity of U.S. shale oil reserves. Energy Econ 67:121–135, ISSN 0140-9883. https://doi.org/10.1016/j.eneco.2017.06.021
47. Demirbas A, Al-Sasi BO, Nizami A-S (2017) Recent volatility in the price of crude oil. Energy Sources Part B 12(5):408–414. https://doi.org/10.1080/15567249.2016.1153751
48. Johnson B (2019) Carbon nation: fossil fuels in the making of American culture. University Press of Kansas, ISBN, p 9780700625819
49. Prăvălie R, Bandoc G (2018) Nuclear energy: between global electricity demand, worldwide decarbonisation imperativeness, and planetary environmental implications. J Environ Manage 209:81–92, ISSN 0301-4797. https://doi.org/10.1016/j.jenvman.2017.12.043
50. Lowry J (2017) Energy produced and carbon released from fossil fuels and the amount of alternative energy required as a replacement. In: Avoiding carbon apocalypse through alternative *energy*. Springer, Cham. https://doi.org/10.1007/978-3-319-52195-4_2
51. Harris TM, Devkota JP, Khanna V, Eranki PL, Landis AE (2018) Logistic growth curve modeling of US energy production and consumption. Renew Sustain Energy Rev 96:46–57, ISSN 1364-0321. https://doi.org/10.1016/j.rser.2018.07.049
52. Shalaeva DS et al (2020) Analysis of the development of global energy production and consumption by fuel type in various regions of the world. IOP Conf Ser: Mater Sci Eng 952:012025
53. Coglianese J, Gerarden TD, Stock JH (2020) The effects of fuel prices, environmental regulations, and other factors on U.S. coal production, 2008–2016. Int Assoc Energy Econ. https://doi.org/10.5547/01956574.41.1.jcog
54. Feijoo F et al (2018) The future of natural gas infrastructure development in the United states. Appl Energy 228:149–166, ISSN 0306-2619. https://doi.org/10.1016/j.apenergy.2018.06.037
55. Jin T, Kim J (2018) What is better for mitigating carbon emissions—renewable energy or nuclear energy? A panel data analysis. Renew Sustain Energy Rev 91:464-471, ISSN 1364-0321. https://doi.org/10.1016/j.rser.2018.04.022
56. Kelly C, Onat NC, Tatari O (2019) Water and carbon footprint reduction potential of renewable energy in the United States: a policy analysis using system dynamics. J Cleaner Prod 228:910–926, ISSN 0959-6526. https://doi.org/10.1016/j.jclepro.2019.04.268

57. Best R, Burke PJ (2018) Adoption of solar and wind energy: the roles of carbon pricing and aggregate policy support. Energy Policy, 118:404–417, ISSN 0301-4215. https://doi.org/10.1016/j.enpol.2018.03.050
58. Dhar A, Anne Naeth M, Dev Jennings P, El-Din MG (2020) Perspectives on environmental impacts and a land reclamation strategy for solar and wind energy systems. Sci Total Environ 718:134602, ISSN 0048-9697. https://doi.org/10.1016/j.scitotenv.2019.134602
59. Dogan E, Ozturk I (2017) The influence of renewable and non-renewable energy consumption and real income on CO_2 emissions in the USA: evidence from structural break tests. Environ Sci Pollut Res 24:10846–10854. https://doi.org/10.1007/s11356-017-8786-y
60. Sources of Greenhouse Gas Emissions, United States Environmental Protection Agency. https://www.epa.gov/ghgemissions/sources-greenhouse-gas-emissions
61. Russel DR, Berger B (2020) US and China climate goals
62. Aiken JD (2021) The Biden climate plan
63. Fitzgerald T, Hassett K, Kallen C, Mulligan CB (2020) An analysis of vice president Biden's economic agenda: the long run impacts of its regulation, taxes, and spending. University of Chicago, Becker Friedman Institute for Economics Working Paper, (2020-157)
64. Clevenger T, Lashof D, 7 Ways the Biden administration can reverse climate rollbacks
65. Maniatis K, Chiaramonti D, van den Heuvel E (2021) Post COVID-19 recovery and 2050 climate change targets: changing the emphasis from promotion of renewables to mandated curtailment of fossil fuels in the EU policies. Energies 14(5):1347
66. Galvin R, Healy N (2020) The green new deal in the United States: what it is and how to pay for it. Energy Res Soc Sci 67:101529
67. Bloomfield J, Steward F (2020) The politics of the green new deal. Pol Q 91(4):770–779
68. South D, Vangala S, Hung K (2021) The Biden administration's approach to addressing climate change. Clim Energy 37(9):8–18
69. Pielke Jr R, Lane N (2020) Memo for president Biden: five steps to getting more from science
70. Roach JA (2021) Exclusive economic zone. In: Excessive maritime claims. Brill Nijhoff, pp 154–175
71. Gruenberg M (2021) Biden makes it official: no keystone XL oil pipeline. Guardian (Sydney) 1947:11
72. Oxford Analytica (2021) US-Canada keystone XL pipeline project is abandoned. Emerald Expert Briefings, (oxan-es)
73. Cartwright ED (2021) Climate and clean energy to take center stage
74. Choi H (2021) President Biden and climate change: policy and issues. Available at SSRN 3847183
75. Windpower Engineering & Development. Wind power was the largest U.S. renewable energy provider in 2019. April 16, 2020. https://www.windpowerengineering.com/wind-power-was-the-largest-u-s-renewable-energy-provider/
76. American Clean Power. 2021 Clean Energy Labor Supply. https://cleanpower.org/wp-content/uploads/2021/06/ACP-Labor-Supply-Report.pdf
77. NS Energy. https://www.nsenergybusiness.com/
78. American clean power. https://cleanpower.org/
79. WindPower, Engineering and Development. https://www.windpowerengineering.com/
80. NS Energy Staff Writer. Top five US states for wind power production profiled. 02 Apr 2020. https://www.nsenergybusiness.com/features/wind-power-states-us/
81. Stromsta K-E (2019) New Jersey more than doubles offshore wind target to 7.5 GW. 19 Nov 2019. https://www.greentechmedia.com/articles/read/new-jersey-governor-more-than-doubles-offshore-wind-target-to-7-5gw
82. Office of Energy Efficiency & Renewable Energy. Computing America's Offshore Wind Energy Potential. 9 Sept 2016. https://www.energy.gov/eere/articles/computing-america-s-offshore-wind-energy-potential
83. Crooks E. The disruptive potential of US offshore wind. 31 July 2020. https://www.woodmac.com/news/opinion/the-disruptive-potential-of-us-offshore-wind/
84. Insight, S. U. S. M. (2019) Year in review; 2020

85. Renewable Energy World. U.S. solar saw record growth during the pandemic year. 3.16.2021. https://www.renewableenergyworld.com/solar/u-s-solar-saw-record-growth-during-the-pandemic-year/#gref
86. Institute for Energy Economics and Financial Analysis. Wood Mackenzie: U.S. will install 324 GW of solar capacity in coming 10 years. March 16, 2021. https://ieefa.org/wood-mackenzie-u-s-will-install-324gw-of-solar-capacity-in-coming-10-years/
87. SEIA, Wood mackenzie power & renewables. Solar market insight report 2021 Q2. 15 June 2021. https://www.seia.org/research-resources/solar-market-insight-report-2021-q2
88. Shaping the grid of the future, FERC. https://www.ferc.gov/
89. Wagman D, PV magazine. About 9.7 GW of large-scale PV set for 2021 delivery in US market. 29 Dec 2020. https://www.pv-magazine.com/2020/12/29/about-9-7-gw-of-large-scale-pv-set-for-2021-delivery-in-us-market/
90. Misbrener K, Solar power world. Hecate Energy commences construction on 600-MW solar portfolio in Texas and Ohio. 21 May 2021. https://www.solarpowerworldonline.com/2021/05/hecate-energy-commences-construction-600-mw-solar-texas-ohio/
91. Boretti A, Castelletto S (2021) Cost and performance of CSP and PV plants of capacity above 100 MW operating in the United States of America. Renew Energy Focus
92. Canadian Solar Industries Association (CanSIA) www.cansia.ca. Natural Resources Canada/Government of Canada
93. Powering up Nevada a Report on the Economic Benefits of Renewable Electricity Development. A Renewable America. Jan 2015. https://energy.nv.gov/uploadedFiles/energynvgov/content/Media/DGA_Report_NV_v5_SMALL.PDF
94. Venables M (2008) The bigger-[power solar]. Eng Technol 3(18):52–53
95. https://www.energy.gov/lpo/ivanpah
96. Ho CK, Sims CA, Christian JM (2015) Evaluation of glare at the Ivanpah solar electric generating system. Energy Proc 69:1296–1305
97. Rib DM (1995) Environmental aspects of SEGS solar power plants (No. CONF-950725-). American Solar Energy Society, Boulder, CO (United States)
98. County M, Infrastructure, A. S. Solana generating station
99. Gallego AJ, Fele F, Camacho EF, Yebra L (2013) Observer-based model predictive control of a parabolic-trough field. Sol Energy 97:426–435
100. https://www.abengoa.com/web/en/novedades/mojave/noticias/
101. To, W. Solar power in the United States
102. https://www.energy.gov/lpo/genesis

www.ingramcontent.com/pod-product-compliance
Ingram Content Group UK Ltd.
Pitfield, Milton Keynes, MK11 3LW, UK
UKHW020803290926
15470UKWH00003B/932
* 9 7 8 3 0 3 0 9 2 0 0 9 8 *